The Energetics of Mangrove Forests

Daniel M. Alongi

The Energetics of Mangrove Forests

Springer

Daniel M. Alongi
Australian Institute of Marine Science
Townsville, Queensland
Australia

ISBN 978-1-4020-4270-6 e-ISBN 978-1-4020-4271-3

Library of Congress Control Number: 2008939501

Cover photo: The mixed *Rhizophora* forests on Hinchinbrook Island in Queensland, Australia, the focus of long-term research by the Australian Institute of Marine Science. Photo taken by Fiona Alongi

Printed on acid-free paper

springer.com

For My Young Ladies,
Fiona, Morrigan, and Ligeia

Preface

This book is about processes within mangrove forests that enable mangrove trees and other organisms to grow and decay, soils to accumulate and erode, gases to enter and leave, and solutes and solids to exchange with adjacent coastal waters. Simply, it concerns the study of the physiology and metabolism of the forest, from individual leaves up to the entire ecosystem, including waterways. Viewed in the not-so-distant past as miasmic swamps, mangroves are now viewed by most people as wetlands worth conserving, being both a valuable ecological and economic resource. Nevertheless, despite the change in perception, mangrove destruction continues unabated into the twenty-first century at an average annual rate of about 1–2%; the greatest threats are urban development, aquaculture, and overexploitation of fisheries. With increasing numbers of people living along tropical coasts, the pressure on mangrove resources can be expected to rise for the foreseeable future.

Why another mangrove book? A number of books and reviews of mangroves are currently available. In fact, a perusal of the current scientific literature indicates that research on the ecology and systematics of mangrove forests is steaming ahead at a rapid pace. Most of this literature deals with various aspects of the biology and structure of mangrove forests and their food webs: systematics, floristics, community structure, species distribution, and biodiversity, and how mangrove species composition varies in response to gradients in physical factors, such as salinity and soil type. While such information is extremely important in understanding mangrove forests, a detailed treatment devoted exclusively to the ecological energetics of mangrove forests is lacking. From individuals to ecosystems, a process-oriented approach to understanding mangroves is necessary if a complete picture of them is to emerge to help facilitate their survival and to use their considerable resources in a sustainable way. With predicted changes in climate and sea-level due to global warming, time is running out to understand how these remarkable and beautiful forests live and cope with changes in an increasingly harsh environment. Wise management requires an understanding not just of structure but of ecosystem function as well. If this book makes ecosystem managers and other concerned parties more aware and appreciative of the intricate complexities of how these ecosystems function, it will have achieved its purpose.

Thanks are due to Steven Hall and Ian Poiner, Directors of the Australian Institute of Marine Science during the production of this book, for their encouragement and support. I thank Britta Schaffelke for keeping vigilant eyes on administrative reins; Joanna Ruxton and Mary Ann Temby, our librarians, for patiently searching and finding often obscure references; and to my wife Fiona and to Tim Simmonds for their very fine graphics. Finally, I thank my friends, Cathy Lovelock, Eric Wolanski, and Dave McKinnon, for their thorough reviews and many helpful comments.

Australian Institute of Marine Science
Townsville, Queensland, Australia

Daniel M. Alongi

Contents

"Among the forms of plants which are sure to attract attention in the tropics are the mangroves, which grow between tide-marks on coasts and estuaries. ...These trees greatly aid the formation of new land, as the mass of aerial roots which arch out from the stem to a considerable distance collects mud and floating refuse, and so raises and consolidates the shore; while the young plants often dropping from the farthest extremity of the branches, rapidly extend the domain of vegetation to the farthest possible limits. The branches, too, send down slender roots like those of the banyan, and become independent trees. Thus a woody labyrinth is formed; and the network of tough roots and stems resists the action of the tides, and enables the mud brought down from the great tropical rivers to be converted into solid land far more rapidly than it could be without this aid...."

(Alfred Russel Wallace, 1878, *Tropical Nature and Other Essays*)

Chapter 1
Introduction

Forests inhabit the edge of tropical seas. As cogently noted by Alfred Russel Wallace during his equatorial travels, mangrove forests are crucial occupiers of the boundary between land and sea, being key ecosystems along many tropical and subtropical coastlines. Mangrove forests are true ecotones, being not just transitional in nature, having some elements of terrestrial and marine ecosystems, but having ecological characteristics all their own.

Mangrove forests are architecturally simple, usually harboring few tree species and lacking an understory of ferns and shrubs, compared with tropical terrestrial forests. Despite their apparent simplicity, however, the standing biomass of mangrove forests can be immense, especially in equatorial regions, rivalling the biomass of many tropical rainforests.

Mangrove trees grow in subtropical and tropical latitudes, with their global distribution best delimited by major ocean currents and the 20°C isotherm of seawater in winter (Fig. 1.1). There are 9 orders, 20 families, 27 genera, and roughly 70 species of mangrove. The most diverse region is the Indo-West Pacific (Fig. 1.1). Indonesia, Australia, Brazil, and Nigeria accommodate roughly 43% of the world's mangrove forests, which comprise a total area of approximately 160,000 km^2 (FAO 2003).

Mangroves are limited globally by temperature, but variations in rainfall, tides, waves, and river flow greatly determine the expanse and biomass of mangroves at the regional and local scale. Several attempts have been made to classify mangroves, but in reality, most forests represent a continuum of morphological types based on broader, river-dominated, tide-dominated, drowned river valley, and carbonate geophysical settings (Woodroffe 1992). Many physical and biological variations are often expressed within a single estuary, as variations in waves, tides, river flow, and rainfall affect water circulation by generating advective and longitudinal mixing and trapping of coastal water (Wolanski 1992). The turbulence induced by these temporal and spatial changes influences the rate of erosion and deposition of soil on which mangroves colonize and grow.

For an individual tree, multiple factors operate to control photosynthesis and growth, including solar radiation, temperature, and the availability of oxygen, nutrients, and water (Ball 1988, 1996). Mangroves are typically distributed from mean sea-level to highest spring tide, with the most conspicuous feature of local forests being the sequential change in tree species parallel to shore. Gradients in salinity,

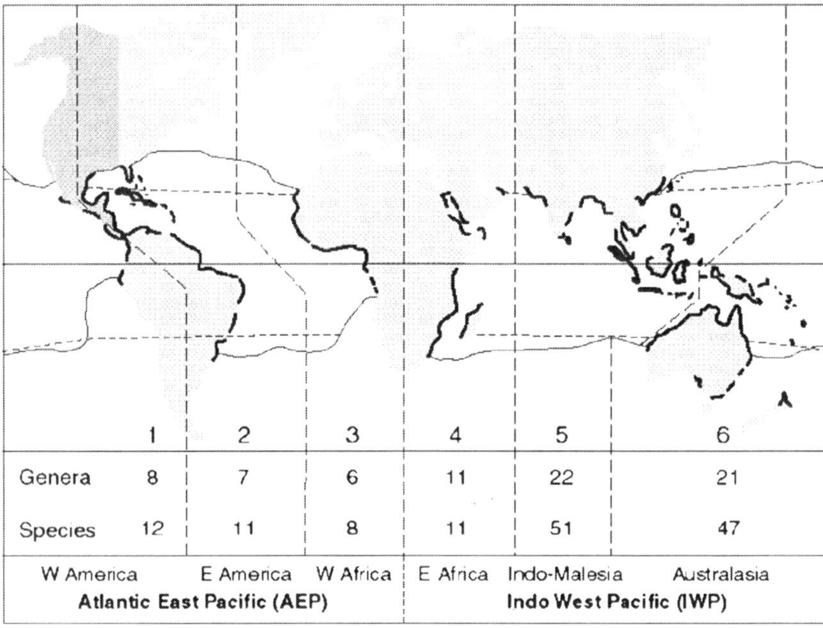

Fig. 1.1 Global distribution of the world's mangrove forests and their biogeographic provinces. Forests are designated as heavy lines. The number of genera and species within each province are noted below the map (Modified from Spalding et al. 1997 and Duke et al. 1998)

soil type and chemistry, nutrient content, physiological tolerances, predation, and competition have all been cited as important factors controlling the zonation of mangroves. In truth, a combination of factors (or better, different combinations of factors!) comes into play over space and time to shape the intertidal distribution of mangroves (Ball 1996; Bunt 1996). To complicate matters, different time scales operate on different parts of mangrove ecosystems; changes in microbial and physiological processes operate on a scale of seconds to hours, trophic interactions from minutes to months, tree growth and replacement on the scale of months to years, and regional changes in forest succession on the order of years to decades.

Against a backdrop of natural change lies the problem of human disturbance. Our ability to discriminate one from the other is usually limited as natural and human disturbances are often indistinguishable and intertwined (Alongi 2008; Piou et al. 2008). Cyclones, lightning, tsunami, floods, disease, and pests naturally disturb mangrove forests, which become even more susceptible when stressed by the introduction of pollutants such as hydrocarbons, herbicides, metals, pesticides, sewage, and acids. Mangroves exhibit considerable resilience to disturbance, coming and going speedily on a geological timeframe, and undergoing perpetual change in forest and ecosystem development commensurate with the morphological evolution of shorelines (Woodroffe 1992; Berger et al. 2006). Mangrove forests are mosaics of interrupted or arrested successional stages, the end result of varying natural responses, such as competition and dispersal, against a background history of disturbance (Berger et al. 2006; Alongi 2008; Piou et al. 2008).

Despite their natural resilience, approximately 50% of the world's mangrove forests have been lost over the past half century (Alongi 2002). The major causes for this destruction have been urban development, aquaculture, mining, and overexploitation for timber, fish, crustaceans, and shellfish. Ironically, many losses reflect the fact that mangroves are a valuable economic resource, being important breeding grounds and nursery sites for birds, fish, crustaceans, amphibians, shellfish, reptiles, and mammals; a potentially renewable resource of wood; and accumulation sites for sediment, carbon, contaminants, and nutrients. Mangroves also offer some protection against coastal erosion and catastrophic events, such as tsunami.

Given the economic imperatives of developing countries, most immediate value and emphasis is placed on fish and wood production. The average monetary value of mangroves has been estimated at US\$10,000 ha^{-1} $year^{-1}$, second only to the values of estuaries and seagrass beds, and greater than the economic value of coral reefs, continental shelves, and the open sea. Globally, the world's mangrove forest resources are worth an estimated US\$180,900,000,000, based on the valuation of Costanza et al. (1998).

The competing demands of humans and mangroves are manageable if relevant scientific information is collected and used to design proper management plans. An ecosystems approach that considers and integrates the relationships between organisms, the physical environment, and humans, can offer crucial insight into the functional role of mangrove forests in the coastal zone, including estimates of maximum sustainable yield of fish and wood.

As in all forested ecosystems, mangrove trees are powered by the sun, the ultimate energy source for the biosphere. Other energy sources are important, for example, wind, lightning, and tides, but the production of trees is directly generated by the sun's light energy. Energy also flows out of the ecosystem in the form of heat (respiration) and in other transformed or processed forms, such as detritus. Being necessary for all life, water, gases, and nutrients constantly enter and leave; so do some biota (fish, plankton, seeds, spores, etc.). These ecosystem components are illustrated in a simple ecosystem model, emphasizing the flow of energy and materials (Fig. 1.2).

Unlike nutrients and water, energy cannot be re-used. Being a unidirectional flow that can be transformed by the biological community (e.g., converted to organic matter), most energy passes out of the ecosystem as heat. The biological community is functionally composed of (1) autotrophs (trees, phytoplankton, benthic algae) which fix the sun's light energy and use simple inorganic compounds to make complex organic structures; and (2) heterotrophs, which use, transform, and decompose the organic matter fixed by autotrophs and imported into the system (Falkowski et al. 2008). This 'producer-consumer' concept is, of course, overly simplistic, but usually represents the bulk of energy and material flow through most ecosystems, including mangroves.

Environmental factors and their gradients across time and space are important forces driving any ecosystem. Water, carbon dioxide and other gases, and inorganic

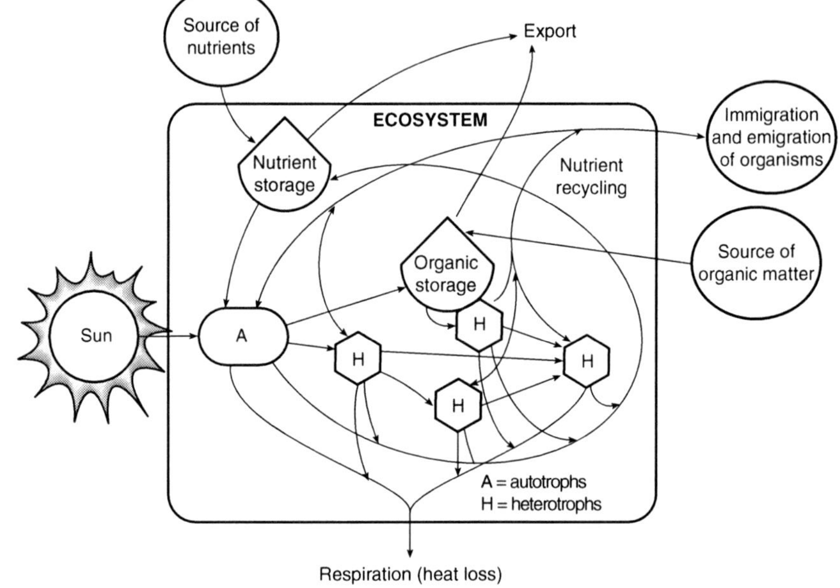

Fig. 1.2 Conceptual model of the flow of energy and materials through an idealized ecosystem

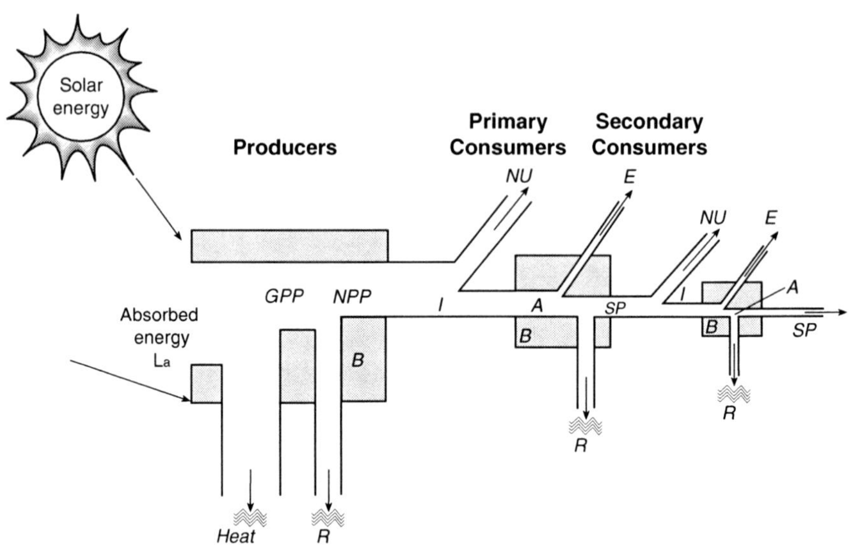

Fig. 1.3 Conceptual model of the attenuation of energy flow from primary producers to higher consumers (Modified from Odum and Barrett 2005). GPP = gross primary production, NPP = net primary production, R = respiration, B = biomass, La = light absorbed by plants, I = ingestion, A = assimilation, SP = secondary production, E = egestion or unassimilated energy, NU = unconsumed energy

and organic solutes, such as oxygen, nitrogen, calcium, sulfur, phosphorus, and amino acids, are important players in regulating ecosystems, but only a small measure of these essential nutrients are immediately available to organisms, as most are stored in less easily extractable forms in soil and water.

The rate of release of essential elements and compounds, solar radiation, and changes in daylength, temperature, and other climatic conditions, are the most important processes regulating ecosystem function; to these we can add tides and changes in salt concentration as important regulatory forces for mangroves. Ultimately, the ecological energetics of all living processes is constrained by the laws of thermodynamics (law of conservation of energy, law of entropy). Ecosystems and organisms are open, non-equilibrium systems that continuously exchange energy and matter with the environment to decrease internal entropy and to increase external energy (Odum and Barrett 2005). Entropy can be most easily envisioned as 'disorder' (low entropy = a low amount of disorder) where order is maintained in a living organism or ecosystem by total community respiration, that is, dissipation of heat. This explains why there is attenuation of energy (and production) from primary producers to secondary consumers (Fig. 1.3).

In the next chapter, we begin with the most obvious and dominant feature of mangrove forests—the trees and their canopy—focusing on how they grow and produce biomass while coping with a harsh tropical, saline and often anoxic environment. Subsequent chapters deal with life in tidal waters in relation to the complexities of water motion, processes occurring on and beneath the forest floor, and finally, scaling up to consider the ecosystem dynamics of mangroves.

Chapter 2
Trees and Canopies

2.1 Introduction

Mangrove trees and forests have obvious similarities to their terrestrial counterparts, yet other attributes of mangroves, especially physiological and morphological specializations, make them structurally and functionally unique. These attributes include:

- Aerial roots
- Viviparous embryos
- Tidal dispersal of propagules
- Frequent absence of an understory
- Frequent absence of annual growth rings
- Highly efficient mechanisms for nutrient retention
- Physiological mechanisms to tolerate salt

Because of the high irradiances and temperatures that occur in the tropics, one would expect that mangroves are C_4 (CAM) plants. However, isotopic ^{13}C signatures for a number of species show only C_3 carbon fixation (Andrews and Muller 1985). Some work suggested a modified C_4 pathway in *Rhizophora*, but all physiological studies indicate that mangroves obtain water, prevent desiccation, and closely coordinate stomatal conductance and CO_2 assimilation, without use of a C_4 pathway (Joshi et al. 1984; Martin and Loeschen 1993). Other C_3 characteristics include a temperature-dependent CO_2 compensation point of 40–90 μl^{-1}, and a temperature optimum for photosynthesis below 35°C (see review of Clough 1992).

Mangrove trees have architecture that allows for efficient light interception and stability for growth on soft soils. An architectural analysis of *Rhizophora* trees indicates unique patterns of tree expansion based on prop root development, drop roots from branches, and multiple stems, all of which result in a dense canopy with little or no understory (Halle et al. 1978; Tomlinson 1986). Thick evergreen leaves and the development of a tap root limited to seedling germination, but surpassed by lateral and adventitious roots, are characteristics of mangroves. Anatomical adaptations include complex leaf anatomy with conspicuous xerophytic and halophytic features: specialized epidermal cells with glandular hairs that function in salt secretion; the leaves of some species contain tannin-containing cells in the hypodermis;

and calcium oxalate druses (Roth 1992). Various specialized cells exist in different species, especially in the aerial roots (Tomlinson 1986).

Most mangroves have flowers and are hermaphroditic with species-specific variations in flower and pollen structure. Pollination is done almost completely by canopy dwellers such as bats, birds, moths, butterflies, bees and other insects. Various types of germination occur, as with terrestrial plants, including a unique type common among mangrove species in which the cotyledons remain in the fruit and the root tip and hypocotyl are exposed. Vivipary (true vivipary to crypto-vivpary) is exhibited by nearly one-third of all species. The advantage of vivipary is that it presumably allows for rapid establishment of seedlings (Krauss et al. 2008). Presumably for this reason, most mangroves appear to be self-fertile. Mangrove propagules are buoyant, dispersing usually but not always at the end of the wet season. A complete description of mangrove reproductive biology and anatomy can be found in Tomlinson (1986), Saenger (2002) and Krauss et al. (2008).

2.2 Biomass Allocation

2.2.1 Partitioning of Tree Components

Like other trees, the partitioning of photosynthetically fixed carbon for growth among different mangrove tree parts varies with species and tree age. Other factors also come into play, such as salinity (Clough 1992), making accurate measurement of biomass important to develop an understanding of tree growth and stand dynamics.

Analysis of allometric data for mangrove trees of different size gives some idea of how photosynthate is allocated for the growth of roots, stems, branches, and leaves (Komiyama et al. 2008). In the case of *Rhizophora* species, the bulk of the biomass is vested in trunks with less fixed carbon vested in prop roots, branches, and leaves (Clough 1992). As the size of a *Rhizophora* tree increases, there is an increasing allocation to above-ground prop roots at the expense of stem weight, presumably a reflection of the fact that larger-sized trees require more support (Fig. 2.1).

In the absence of reliable measurements of respiratory losses from the prop roots and stem, the accumulation of biomass in these tree structures gives a crude estimate of how fixed carbon is allocated between them as the tree grows. In contrast, flowers, leaves, twigs, and branches are shed as litter throughout the year, so carbon must be allocated to replace them in addition to that allocated for growth and expansion of the canopy (Clough 1992).

A comparison of the partitioning of biomass of the same species in contrasting environments gives us considerable insight into how mangrove trees respond to environmental stress over time. In such a comparison between *Rhizophora stylosa* trees from the wet tropics of Queensland, Australia and those living along the arid coast of Western Australia (Fig. 2.1), Clough et al. (1997a) found significant differences in

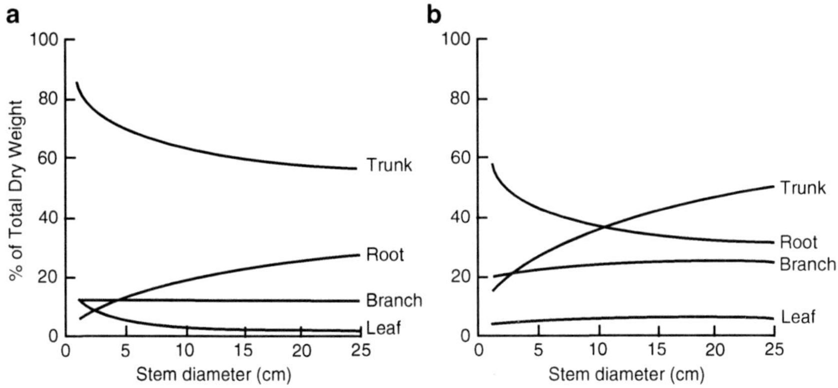

Fig. 2.1 Partitioning of above-ground biomass in relation to tree size for *Rhizophora stylosa* of (**a**) wet tropical Queensland and (**b**) arid Western Australia (Modified from Clough et al. 1997a)

biomass partitioning. For wet tropical trees, the contribution of the trunk (stem + butt) to total dry weight decreased from 71% to 57% from a stem diameter of 5 cm up to 25 cm. The contribution of prop roots conversely increased over the same stem diameter from 12% to 28% of total dry weight. In contrast, in arid-zone trees, the contribution of the trunk increased from 25% to 50% over the 5–25 cm stem diameter range, with a decrease (40–30%) in the contribution of prop roots to total tree dry weight. The most plausible explanation for these differences is that the water relations of mangroves are contrasting between wet and dry tropical environments. In arid Western Australia, soil salinities are high and are likely to negatively affect water use (Passioura et al. 1992; Ball and Passioura 1995). Under such conditions, it may be advantageous for the trees to invest more fixed carbon in growing more extensive root systems to maximize water gain.

A positive relationship would be expected between above- and below-ground root biomass in the Australian study, but carbon allocation for the production of below-ground roots is much more difficult to estimate as roots in the soil comprise soft, non-woody, lateral roots of <10 mm diameter and fine fibrous roots <1 mm in diameter (Gong and Ong 1990; Robertson and Dixon 1993). Below-ground roots may only represent about 10–15% of total tree biomass (Gong and Ong 1990; Alongi et al. 2003a), but the allocation of fixed carbon to replace sloughed root hairs and roots is considerably greater (McKee and Faulkner 2000; Cahoon et al. 2003; Sánchez 2005).

Application of a method using silica gel to distinguish dead from live fine roots (Robertson and Dixon 1993; Gleason and Ewel 2002) has repeatedly found that most fine roots are dead, probably a reflection of rapid root turnover coupled with slow rates of decomposition (McKee 2001). Vertical profiles of fine roots (Fig. 2.2) show not only the proliferation of dead roots but, more importantly, that live roots are usually most abundant in the upper soil layers (0–40 cm) and can be found to at least a soil depth of about 1 m.

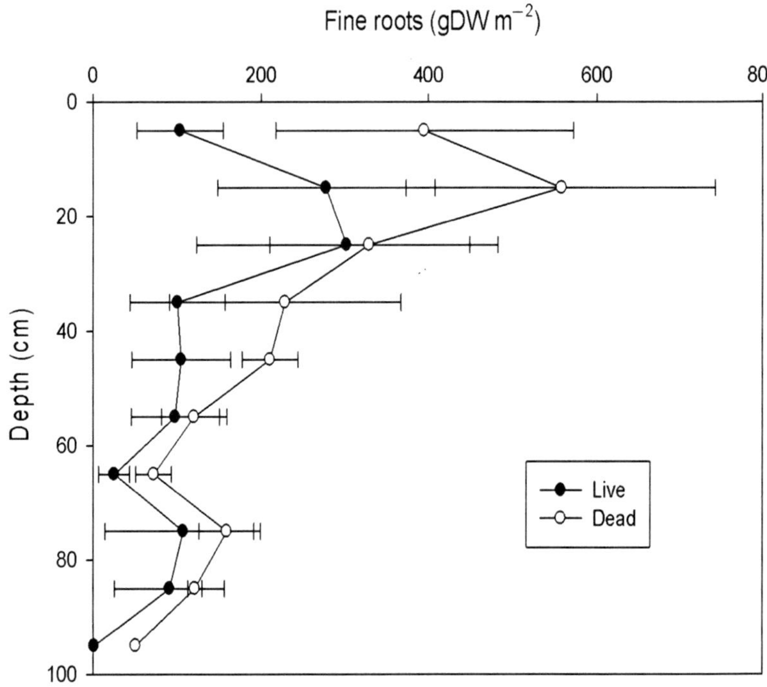

Fig. 2.2 Vertical soil profiles of live and dead fine roots in a low intertidal *Kandelia candel* forest of southern China (Modified from Alongi et al. 2005b)

The preponderance of dead root biomass in mangroves can serve as a nutrient conserving mechanism. Even large dead roots may serve this purpose. In old root channels formed by the mangroves *Rhizophora mangle* and *Avicennia germinans* in central Belize, McKee (2001) found that roots proliferated in decaying roots and channels taking paths of least resistance and recovering nutrients released from decomposing roots.

Estimates of below-ground biomass are rare for mangrove forests (Komiyama et al. 1987, 2008; Matsui 1998), but some workers have shown that mangroves have a higher below- to above-ground biomass ratio than for other vegetation (Saenger 1982; Snedaker 1995; Sánchez 2005), supporting the idea that mangroves allocate much of their fixed carbon to roots. However, recent studies differentiating live from dead roots tell a different story (Table 2.1). Averaging all species and ages, the overall below- to above-ground biomass ratio is 19% (range = 7–55%). A study of South African mangroves similarly found a low (9%) ratio, although live and dead roots were not distinguished (Steinke et al. 1995). An analysis of tropical forest biomass data (Fittkau and Klinge 1973; Golley et al. 1975; Proctor 1987; Medina and Cuevas 1989; Yamada 1997, Barnes et al. 1998; Clark et al. 2001) results in an overall mean ratio of 18% (range = 3–50%). Given the wide range of values, the oft-repeated statement that mangrove forests have higher below- to above-ground biomass ratios cannot be sustained (Komiyama et al. 2008; Lovelock 2008).

Mangroves may differ from other tropical forests in that the proportion of live to dead roots tends to be greater in terrestrial forests (Murach et al. 1998), although the overlap in values between ecosystems is probably great.

Table 2.1 also suggests that there may be species-specific differences in the allocation of below- and above-ground biomass. Moreover, there are clear species differences in the distribution of biomass among tree components, especially considering species differences in architecture (Fig. 2.3). Obviously, *Rhizophora* species vest a considerable proportion of fixed carbon in prop roots compared to species of contrasting architecture. Nevertheless, there are regional variations within the same species (Alongi et al. 2003a) that may be the result of environmental, and perhaps genetic, differences.

As in other forests, there is a shift in the proportion of biomass among tree components with increasing age (Fromard et al. 1998; Matsui 1998). The proportion of fixed carbon vested in leaves shows a small decline as more biomass is shifted to trunks (Fig. 2.4) reflecting growth in the size of the tree. With increasing size (and age), the ratio of below- to above-ground biomass increases, at least for young stands of *Rhizophora stylosa* (Fig. 2.5). The data are few, but do point to the fact that mangroves partition biomass among various tree components similarly to other tropical trees (Turner 2001; Lovelock 2008).

Table 2.1 Above- and below-ground biomass (t DW ha^{-1}) in different mangrove forests throughout Asia and Australia. Forest ages are given where known (Data taken from Alongi and Dixon 2000; Alongi et al. 2000a, 2003a, 2004a, b, 2005b; Clough 1998; Matsui 1998; Alongi and Clough unpublished data)

Species/location	AGB	LBGB	DBGB	LBGB/AGB ratio (%)
RS/WA	246.7	44.8	251.2	18
RS/WA	282.8	55.8	104.4	20
RS/WA	207.9	36.3	153.0	17
AM/WA	45.8	21.2	201.3	46
AM/WA	147.6	11.5	91.7	8
AM/WA	90.5	16.1	366.0	18
RS, RA/QLD	619	53.2	322.2	9
RA/TH (3 years)	65.4	11.2	258.7	17
RA/TH (5 years)	42.0	23.1	117.5	55
RA/TH (25 years)	344.2	35.6	317.1	10
RA/ MA (5 years)	106	9.8	231.8	9
RA/MA (18 years)	352	24.5	143.6	7
RA/MA (85 years)	576	48.1	223.3	8
KC/CH (5 years)	16	4.3	12.6	27
KC/CH (20 years)	93	18.3	34.4	20
KC/CH (30 years)	133	13.9	32.5	10

Abbreviations: AGB = above-ground biomass, LBGB = living below-ground biomass, DBGB = dead below-ground biomass, RS = *Rhizophora stylosa*, WA = Western Australia, AM = *Avicennia marina*, RA = *Rhizophora apiculata*, QLD = Queensland, TH = Thailand, MA = Malaysia, KC = *Kandelia candel*, CH = China

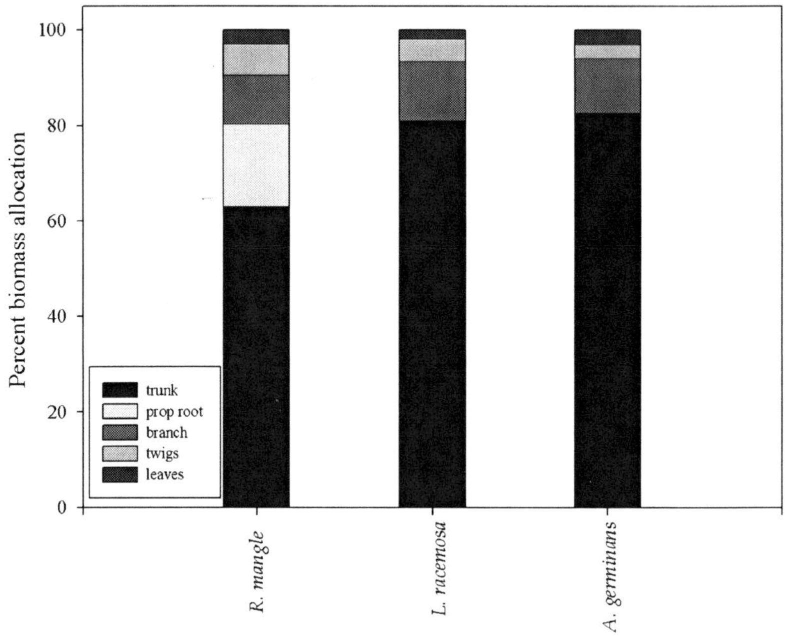

Fig. 2.3 Percentage of dry weight biomass allocated to the components of three mangrove species in the Dominican Republic (Data from Sherman et al. 2003)

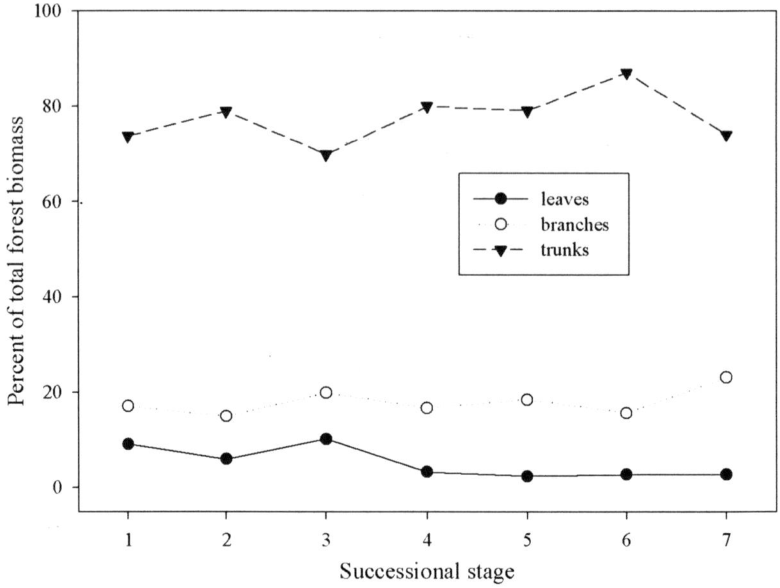

Fig. 2.4 Changes in the partitioning of plant biomass within increasing successional stage in mangrove forests of French Guiana. Stages 1–3 are pioneering and young stages and stages 4–8 are mature stands (Data from Fromard et al. 1998)

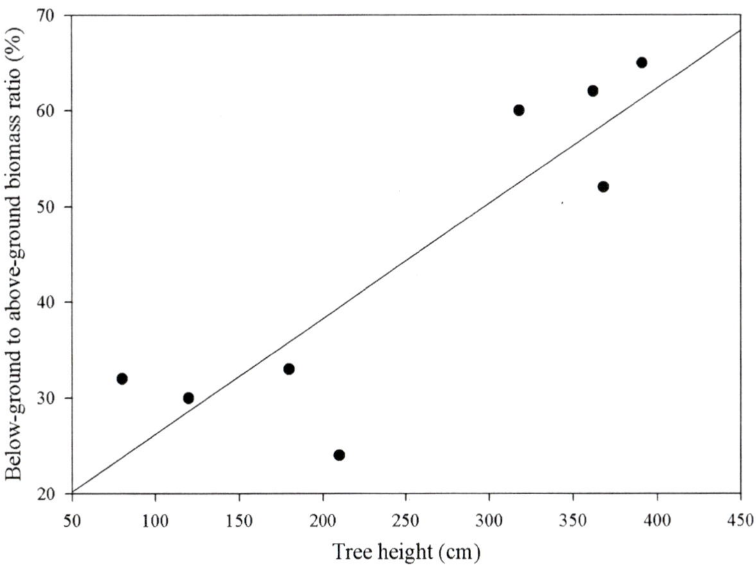

Fig. 2.5 Increase in the ratio of below-ground to above-ground biomass with increasing tree height in young stands of *Rhizophora stylosa* on Iriomote Island, Japan (Data from Matsui 1998)

2.2.2 Global Patterns of Mangrove Biomass

Mangrove forest biomass varies enormously across the globe, with variations within regions caused by many factors, including stand age, species composition, and responses to environmental conditions. Despite huge variability, there is a clear relationship of height of the vegetation and latitude, with both factors causally related to biomass (Saenger and Snedaker 1993). A compilation of known above-ground biomass data shows a trend of declining forest biomass with increasing distance from the equator (Fig. 2.6), reflecting the limitation of mangroves to tropical and subtropical latitudes. Above-ground biomass ranges from 619 t ha^{-1} of mature *Rhizophora* forests in Hinchinbrook Channel, Australia (Clough 1998) to 6.8 t ha^{-1} of small *Avicennia marina* in Tuff Crater, New Zealand (Woodroffe 1985a). Unpublished AIMS data (not included in Fig. 2.6) from the Fly River delta in Papua New Guinea shows values up to 680 t ha^{-1} for tall mixed *Rhizophora* and *Bruguiera* forests. A linear regression of the data yields the equation, $y = 285.979 - 6.043x$ with a significant correlation of 0.188 ($P < 0.001$). The low correlation coefficient no doubt reflects the variance in the data as there are genuine differences in the size and composition of forests in response to stand age, soil texture, nutrient availability, rainfall, salinity, and temperature.

How do mangrove forests compare in terms of above-ground biomass with tropical terrestrial forests? Figure 2.7 shows the median and percentiles of data sets of above-ground biomass for mature upland forests and mangrove forests from

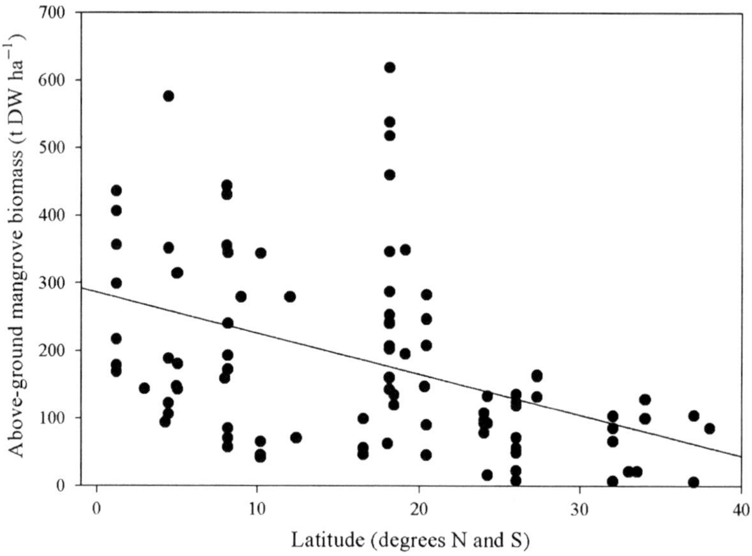

Fig. 2.6 Latitudinal trends in mangrove forest biomass (t DW ha^{-1}). The line represents a linear regression of the data (Updated from Saenger and Snedaker 1993. Using data in Clough 1998; Fromard et al. 1998; Alongi and Dixon 2000; Alongi et al. 2000a, b, 2004a, 2005b; Sherman et al. 2003; Alongi and de Carvalho 2008)

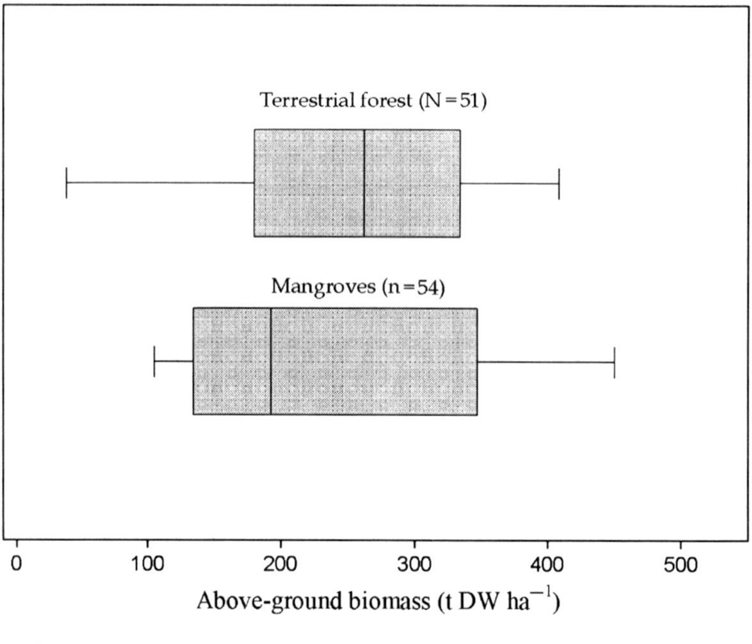

Fig. 2.7 Global comparison of above-ground biomass of terrestrial forests and mangrove forests from the equator to 20° N and S latitude. Mangrove data from sources cited in Fig. 2.6 (Terrestrial data from Fittkau and Klinge 1973; Golley et al. 1975; Grubb and Edwards 1982; Vitousek and Sanford 1986; Medina and Cuevas 1989; Bruenig 1990; Yamada 1997). Vertical line within each box is the median and the ends of the boxes are the 25% and 75% percentiles, respectively.

the equator to 20° N and S latitude. Excluded from this analysis are plantations, immature or pioneering stands, and data from sources where the methodology is unclear or unspecified. The median biomass for tropical terrestrial forests is 262 t DW ha^{-1} with 25% and 75% percentiles of 181 and 347 t DW ha^{-1}, respectively; mean biomass is 246 t ha^{-1}. The median biomass for mangroves is 193 t DW ha^{-1}, with 25% and 75% percentiles of 135 and 347 t DW ha^{-1} with a mean biomass of 247 t DW ha^{-1}. The median biomass of tropical terrestrial forests is greater, but the mean biomass is virtually identical, precluding any clear distinction between the biomass of mangrove and tropical terrestrial forests. It must be remembered that these data are underestimates of true forest biomass as they do not include biomass of below-ground roots.

2.2.3 Nutrient Capital

The allocation of mineral nutrients among the components of mangrove forests has been little studied, although a large number of studies have measured foliar concentrations of many elements in many species. The lack of whole-forest nutrient inventories is unfortunate because knowledge of the allocation of macro- and micronutrients among ecosystem components can offer incisive clues into utilization and storage within forested ecosystems; equivalent information is abundant for tropical terrestrial forests (Vitousek and Sanford 1986; Barnes et al. 1998; Perry et al. 2008).

The sparse data for mangroves show an allocation of elements among tree components and soils similar to those of other tropical forests (Alongi et al. 2003a, 2004b). Among *Rhizophora stylosa* and *Avicennia marina* forests along the arid coast of Western Australia, prop roots and stems are the largest single tree components for both species, respectively, but most nutrients are stored in leaves and living roots of both species (Alongi et al. 2003a). The vast bulk of elements are stored in soils; only a small fraction of the total nutrient pool is stored in tree biomass. A large below-ground pool of dead fine roots was identified at all stands, equivalent to 36–88% of total living tree biomass. For *R. stylosa*, the amount of calcium, sulfur, chloride, sodium, silicon, iron, manganese, zinc, boron, molybdenum, and copper vested in dead roots was greater than in total living tree biomass (Table 2.2); this is true for all elements in the *A. marina* forests. The proportion of iron and manganese in roots was disproportionately large, consistent with evidence of metal plaques on mangrove roots.

How mangroves proportion essential elements in their biomass varies with forest age (Table 2.3). In southern Thailand, Alongi et al. (2004b) examined the partitioning and storage of elements in trees and soils of *Rhizophora apiculata* forests of different age (3, 5, and 25 years). Three patterns emerged with increasing forest age:

- Concentrations of most (but not all) elements in various tree parts declined
- Soil pool size of most elements decreased
- Proportion of C, N, S, Na, Mn, Zn, and Mo in living biomass increased

Table 2.2 Mean inventory of chemical elements (kg ha^{-1}) in living tree biomass, dead fine roots, and soils from *R. stylosa* and *A. marina* forests in arid Western Australia. Values are means of three stands of both species (Adapted from Alongi et al. 2003a)

	R. stylosa tree	Roots	Soil	*A. marina tree*	Roots	Soil
C	114,780	60,133	169,266	55,387	79,167	117,733
N	532	100	10,330	412	168	11,653
P	59	53	2,050	61	66	2,663
Mg	2,410	1,129	74,110	308	1,291	100,330
Ca	1,505	2,086	384,000	1,025	2,457	344,000
S	2,108	10,572	85,330	576	12,346	44,667
K	581	279	30,333	419	1,037	44,000
Cl	4,606	10,342	99,333	1,440	10,929	106,667
Na	3,156	6,765	110,666	990	8,695	126,333
Si	193	269	2,052,000	34	286	1,544,333
Fe	637	4,737	149,667	138	8,063	211,333
Mn	7	13	1,033	2	15	1,500
Zn	1	4	183	1	5	250
B	9	86	280	3	115	366
Mo	2	9	51	1	31	115
Cu	1	2	87	1	5	115

Table 2.3 Percentage of the total element pool (%) vested in total living biomass in 3-, 5- and 25-year old forests of *R. apiculata* in southern Thailand. The remainder of each element pool is stored in soil and dead biomass (Data from Alongi et al. 2004b)

Element	3-year (%)	5-year (%)	25-year (%)
C	10	13	53
N	2	3	12
Mg	0.4	3	3
Ca	6	3	34
S	0.4	0.6	2
K	0.6	0.5	2
Na	1	2	10
P	3	2	13
Mn	1	2	4
Fe	0.06	0.2	0.2
Zn	0.2	0.3	0.8
Cu	0.7	0.3	0.6
Mo	0.6	0.7	2

These data suggest net accumulation with forest maturity. In terrestrial forests, element concentrations in tree parts also decline with age due to declining rates of photosynthesis and tree growth, soil fertility, and nutrient-use efficiency (Golley et al. 1975; Nwoboshi 1984; Folster and Khanna 1997; Wardle et al. 2004). The same reasons can be offered for mangroves.

Do tropical forests store proportionally more nutrients in their biomass than in soils compared with forests of higher latitude? The sparse mangrove data supports the idea that this pattern may not be true for all tropical forests (Grubb 1995; Barnes et al. 1998; Perry et al. 2008). A comparison of mangroves with boreal, temperate, and other tropical forests (Golley et al. 1975; Jordan 1985; Proctor 1989) indicates that mangroves contain a lower percentage of nitrogen in living biomass relative to total ecosystem N than in other forests. The total amount of nitrogen in living tree biomass is similar to that in boreal and temperate forests, but less than in tropical rainforests and savannas. This may imply that mangroves differ from other trees in rates of mineral cycling and mean residence times of essential elements (see Section 6.5).

Foliar concentrations of most elements for most mangrove species (Spain and Holt 1980; Silva et al. 1990; Jayasekera 1991; Thomas and Fernandez 1997; Alongi et al. 2003a, 2004b) are either lower or at the lower end of the range of values reported for other tropical trees (Drechsel and Zech 1991, 1993; Epstein 1999). Moreover, most element concentrations in mangrove leaves are within the same range implying that mangrove species are able to maintain physiological balance of cations and anions regardless of environmental stress. The costs of such maintenance may be high (López-Hoffman et al. 2007).

2.3 Ecophysiology

Mangrove trees physiologically tolerate or avoid two important factors in the intertidal zone—anoxia and salt. There are a number of excellent reviews of mangrove ecophysiology (Ball 1988, 1996; Clough 1992; Lüttge 1997; Saenger 2002), so only a précis will be given here, including some recent developments.

2.3.1 Anoxia

Being waterlogged by tides and accommodating an active microbial flora, mangrove soils are oxygen-deficient below the upper few mm. Oxygen may penetrate deeper soils within the linings of cracks, fissures, and tubes and burrows of benthic animals. Soil anoxia can influence the growth of mangroves in the following ways:

- Without sufficient oxygen, below-ground roots must rely on internal transport of gases to satisfy their oxygen requirements.
- Low redox conditions mean that some elements are more available, and that some are less available, for plant uptake.
- Some microbial metabolites, such as H_2S, reduced iron and manganese compounds, and organic acids, are toxic to plants.

Mangrove roots must cope with short periods of anoxia, as survival and sustained growth of the plants depends on maintaining oxygen levels in the roots. Their shallow nature and the presence of numerous lenticels and extensive aerenchyma facilitate oxygen availability; most species have some structural features (pneumatophores, knee roots, stilt roots, plank roots) to provide root ventilation via atmospheric exposure, at least during low tides (McKee 1993). Some species have above-ground roots with the ability to photosynthesize and thus provide oxygen directly to underground roots (Yabuki et al. 1985; Dromgoole 1988).

The rate at which oxygen is supplied to the roots below-ground exceeds that required to satisfy the respiration of the roots, resulting in diffusion of oxygen from the root to the surrounding soil. This often leads to the development of an oxidized rhizosphere around the roots, as observed for some genera such as *Avicennia* (Andersen and Kristensen 1988). Most evidence is circumstantial, with differences noted in redox potential between soils near roots and those distal to roots or unvegetated (Thibodeau and Nickerson 1986; McKee and Mendelssohn 1987; Youssef and Saenger 1996).

Despite having some morphological and physiological mechanisms to avoid anoxia, mangroves can be adversely affected by waterlogging (Ball 1988, 1996; Farnsworth 2004). Some responses to waterlogging include:

- Decrease in cytokinin export from the roots
- Accumulation of abscissic acid in the leaves
- Stomatal closure
- Rapid leaf senescence and shedding
- Increased foliar sodium
- Retardation of shoot development and elongation
- Reduction in water uptake and transpiration
- Formation of adventitious roots

Anoxia *per se* has little or no effect on the growth of seedlings of a number of mangrove species (Youssef and Saenger 1998). In experiments examining the effect of root anoxia in the presence and absence of various phytotoxins, Youssef and Saenger (1998) found that *Bruguiera gymnorrhiza* and *Hibiscus tiliaceus* seedlings showed reduced carbon assimilation. Some species (*Avicennia marina, Bruguiera gymnorrhiza, Aegiceras corniculatum*) were affected by reduced iron and manganese, but in all five species studied, the presence of sulfide and/or the addition of various mixtures of three phytotoxins, resulted in the complete inhibition of photosynthesis. Under prolonged anoxia, Youssef (1995) found that after 10 days of exposure, there was a significant amount of ethanol accumulation in the roots of *Aegiceras corniculatum, Avicennia marina, Excoecaria agallocha*, and *Rhizophora stylosa*; neither *Bruguiera gymnorrhiza* nor *Hibiscus tiliaceus* showed any signs of ethanol accumulation. Only *Avicennia marina* was able to maintain high levels of ethanol under long-term anoxia. Mangroves utilize a range of biochemical responses to anoxia, including some accumulation of ethanol and subsequent leakage to the surrounding environment; there may be a change in the ethanol and shikimic acid pathways to produce malic acid. Saenger (2002) suggests that this shift is a good adaptive strategy because malic acid is involved in ionic balance and salt uptake.

Modified physiological and morphological changes to root anoxia can take place, as found in *Rhizophora mangle, Avicennia germinans*, and *Laguncularia racemosa* (McKee 1996). In long-term experiments, *Rhizophora mangle* seedlings were least affected, but oxygen concentrations, respiration rates, and extension rates of the roots of *A. germinans* and *L. racemosa* declined significantly. These results demonstrate that interspecific differences exist among mangroves in their ability to deal with soil anoxia.

Exposure to anaerobic conditions reduces tolerance of most mangroves to increases in salinity, probably because the lack of free oxygen interferes with salt exclusion and selectivity for potassium over sodium (Ball 1988, 1996). Potassium is required in relatively high concentrations for protein synthesis and other metabolic processes (Ball et al. 1987).

2.3.2 *Salt*

Mangroves utilize one or more strategies to accommodate the presence of salt, whereby they absorb some sodium and chloride ions but are able to control uptake sufficiently to maintain an acceptable balance of water. Popp (1995) suggests that mangroves have developed mechanisms of salt avoidance and regulation (i.e., salt resistance) coupled with mechanisms of tolerance. The resistance strategies include salt exclusion, extrusion, storage, succulence, compartmentalization and osmoregulation (see Chapter 3 in Saenger (2002) for an extensive treatment of these strategies).

Growth responses to salinity vary greatly, reflecting a wide range of tolerance among species. Some species, such as *Avicennia marina*, do not grow in freshwater and may be obligate halophytes. Others survive well in freshwater and may not have an obligatory requirement for salt beyond trace amounts (Clough 1992). Under natural conditions, mangroves exhibit clear tolerance differences among species (Table 2.4).

High salinities result to physiological responses similar to terrestrial plants experiencing drought, as highly saline soils have low osmotic potential that constrain water relations of mangroves (Ball 1988, 1996; Popp et al. 1993). As solar radiation peaks from late morning to early afternoon, water loss exceeds water uptake with a resultant diurnal change in shoot water potential followed by rapid recovery, as photon flux declines towards evening. A positive water balance and thus photosynthesis can only be maintained if the potentials in the plant are lower than in the soil. Maintaining water balance in high salinity soil presents the problem of trying to take up sufficient inorganic ions to maintain osmotic balance, yet avoiding the adverse effects of high ionic concentrations in the cytoplasm. Ions are stored in the vacuole to presumably minimize physiological damage. Organic solutes are synthesized to maintain osmotic balance in the cytoplasm (Popp 1995). The metabolic cost of making organic solutes is presumably high, as carbon and nutrients used in solute formation is that less available for growth.

2.3.3 *Balancing Carbon Gain and Water Loss*

Mangroves are normally exposed to high solar radiation in the tropics, and the absorption of this light translates into heat energy such that leaves in the canopy are

Table 2.4 Relative tolerances among mangroves to salinity and their relative frequency with tidal zonation (Modified and updated from Clough 1992)

Species	Salt tolerance	Tidal zone
Acanthus ilicifolius	++	L → H
Aegialitis annulata	++++	M → H
Aegiceras corniculatum	+++	L → M
Avicennia germinans	+++++	L → H
Avicennia marina	+++++	L → H
Bruguiera cylindrica	++	M→H
Bruguiera exaristata	+++	M → H
Bruguiera gymnorrhiza	+++	M → H
Bruguiera parviflora	++	M → H
Bruguiera sexangula	+	M → H
Ceriops decandra	++	M → H
Ceriops australis	++++	M → H
Ceriops tagal	++++	M → H
Cynametra iripa	++	H
Diospyros ferrea	+	H
Excoecaria agallocha	+++	M → H
Heritiera littoralis	++	M → H
Kandelia candel	+++	M → H
Lumnitzera littorea	++++	H
Lumnitzera racemosa	++++	H
Nypa fruticans	+	L → M
Osbornia octodonta	+++	H
Rhizophora apiculata	+++	L → M
Rhizophora lamarckii	+++	L → M
Rhizophora mangle	++++	L → M
Rhizophora mucronata	++	L → M
Rhizophora stylosa	++++	L → M
Sonneratia alba	+++	L → M
Sonneratia caseolaris	+	L → M
Xylocarpus granatum	+++	M → H
Xylocarpus mekongensis	+++	M → H

+++++ = very tolerant to + = not tolerant. For tidal levels, H = high, M= mid and L = low intertidal zones

warmer than ambient temperature. The vapor pressure difference (VPD) between the leaves and the environment thus increases, leading to an increase in the rate of water loss. As noted earlier, mangrove leaves possess morphological adaptations to reduce water loss, but these adaptations and reducing the aperture of the stomata often do not offset losses (Clough 1992).

The stomata also control the influx of CO_2 for photosynthesis, so mangroves are faced with the problem of balancing the need to take up CO_2 for carbon fixation with the need to minimize water loss. Consequently, low transpiration rates are imposed on mangroves because of the environmental conditions in which they live. In rainy months, rates of transpiration increase as water availability increases and heat-related problems decline. Low transpiration rates may serve to minimize salt accumulation around the roots; as roots take up more water they draw down salt which is excluded by the roots. Indeed, it appears that there is a limiting rate of transpiration in mangrove forests. Passioura et al. (1992) calculated a limiting rate of 1 mm day^{-1} which is in the range of field data suggesting an evaporation rate of 2 mm day^{-1}. Since well-developed canopies can transpire at the rate of up to 7 mm day^{-1}, it is clear why mangroves have developed mechanisms to minimize water loss and salt gain.

Rates of transpiration vary among species, but rates range from 0.5 to 6.96 mmol m^{-2} s^{-1} (Saenger 2002). Compared with other C_3 plants, mangroves have low transpiration rates and stomatal conductances, but high water-use efficiencies (Ball 1988; Lovelock and Ball 2002). The data support the generalization that mangroves follow a conservative water-use strategy, with greater conservation with increasing stress. With increasing salt tolerance of a species, water-use efficiency increases (Ball 1996), but transpiration rates may be ultimately limited by constraints imposed by the hydraulic architecture of mangrove shoots and leaves (Sobrado 2000).

Plotting data from studies that have measured transpiration, water-use efficiency, and stomatal conductance concurrently (data in Table 3.6 in Saenger 2002), we can see that there are functional tradeoffs between retaining water and gaining CO_2, reflecting the need to minimize water use while maximizing photosynthesis (Fig. 2.8). At increasing rates of transpiration, water-use efficiency declines. Further, low stomatal conductance restricts water loss also limiting CO_2 uptake, but with a high water-use efficiency. There is a positive correlation between transpiration rate and stomatal conductance (Fig. 2.8). In low salinity environments, Clough and Sim (1989) found that stomatal conductance ranges from 79–271 mmol m^{-2} s^{-1} with assimilation rates ranging from 5.8–19.1 µmol CO_2 m^{-2} s^{-1}; lower rates of stomatal conductance and assimilation occur in arid environments with concomittantly lower rates of photosynthesis (Cheeseman et al. 1997).

In arid habitats other environmental stresses, such as high solar radiation, induce photoinhibition. This is most clearly seen in diurnal patterns of transpiration rate. In wet tropical environments, most mangroves show a marked midday maximum with a steady decline until dusk. In arid environments, transpiration rates are markedly at a minimum at midday to avoid water loss, and leaf orientation is modified to minimize high leaf temperatures. Both stomatal conductance and assimilation rate are maximal at leaf temperatures ranging from 25–30 °C, with a rapid decline above 35°C (Ball 1988).

2.4 Tree Photosynthesis and Respiration

Mangroves utilize an advantageous strategy of minimizing water loss and maximizing carbon gain with high water-use efficiency and low transpiration rates to maximize growth, being flexible depending on environmental conditions. For example, some species minimize energy expenditure by opportunistically maximizing growth during rainy seasons or during short periods of freshwater input, and synchronizing reproductive output during these wet periods. Such physiological plasticity is one reason why mangroves are so successful across the intertidal seascape.

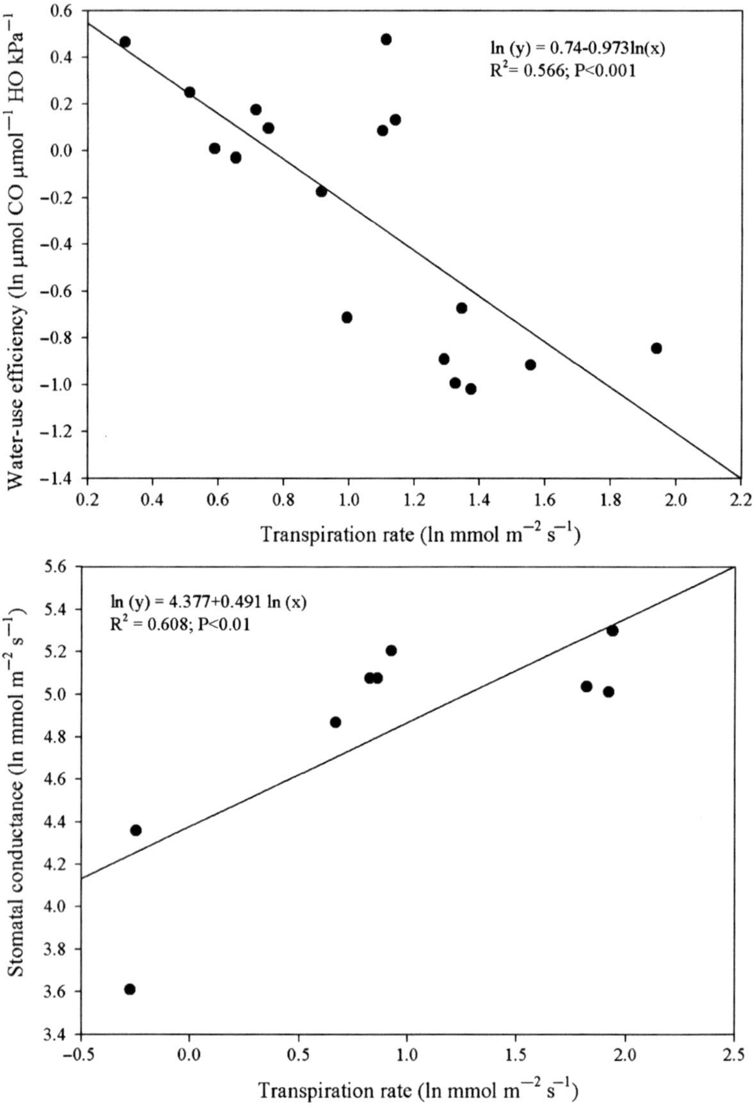

Fig. 2.8 The energetic tradeoff of transpiration rate with water-use efficiency (*top*) and the relationship with stomatal conductance (*bottom*) (Data from Table 3.6 in Saenger 2002)

2.4.1 Photosynthetic Rates

The light response curves of mangroves are similar to other plants, with a steep linear increase up to ≈300–400 μmol photons m^{-2} s^{-1} after which saturation is reached (Fig. 2.9). Under favorable conditions of low vapor pressure

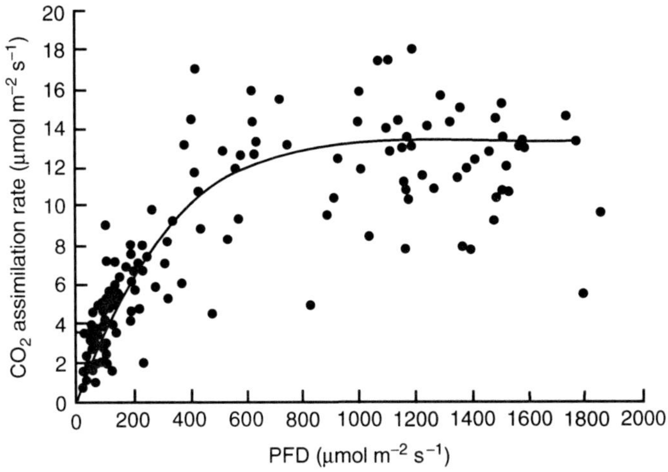

Fig. 2.9 A typical light response curve of a 10 year-old *Rhizophora apiculata* tree, Matang Mangrove Forest Reserve, peninsular Malaysia (Data from Gong et al. 1992)

deficit (<22 mbar) and low salinity (<15), maximum CO_2 assimilation rates may exceed 25 µmol CO_2 m^{-2} s^{-1}, but most rates lie between 5 and 20 µmol CO_2 m^{-2} s^{-1} (Table 2.5). Mangrove photosynthesis reaches saturation at comparatively low light levels due to their low stomatal conductance and intercellular CO_2 concentrations (Clough 1992; Cheeseman 1994; Tuffers et al. 1999; Cheeseman and Lovelock 2004). Clough and Sim (1989) found that for 19 mangrove species in diverse habitats, maximum rates of net photosynthesis under saturated light (>800 µmol photons m^{-2} s^{-1}) decreased with increasing salinity and increasing vapor pressure deficit. *Avicennia marina* had consistently high rates of CO_2 assimilation than *Rhizophora* species which, in turn, had higher assimilation rates than *Bruguiera* species.

Rates of net photosynthesis vary widely among species with the major regulatory factors being soil salinity, vapor pressure deficit between leaf and surrounding air, and light intensity (Lovelock and Ball 2002). Mangrove leaf photosynthesis is constrained by the above-cited factors, but a comparison of CO_2 leaf assimilation rates between mangroves and tropical terrestrial trees (Fig. 2.9) suggests higher median rates of photosynthesis in mangroves. This comparison includes shade-intolerant and shade-tolerant species of terrestrial trees. There is therefore great overlap in rates between and within groups owing to species-specific differences in assimilation rates, position in the canopy, tree age, environmental conditions, and nutrient availability. Among terrestrial trees, shade-intolerant species have a median photosynthetic rate of 13 µmol CO_2 m^{-2} s^{-1}, virtually identical to mangroves (12 µmol CO_2 m^{-2} s^{-1}). Grouping all terrestrial trees, the median rate is 7 µmol CO_2 m^{-2} s^{-1} (Fig. 2.10). One clear difference is that lower values for both groups are most often measured in the dry tropics, underscoring the importance of climate.

Table 2.5 Net rates of maximum light-saturated leaf photosynthesis (A, μmol CO_2 m^{-2} s^{-1}) in mangroves from various locations (Data taken from Miller 1975; Ball et al. 1988; Björkman et al. 1988; Clough and Sim 1989; Smith et al. 1989a; Gong et al. 1992; Lin and Sternberg 1992; Ong et al. 1995; Cheeseman et al. 1997; Clough et al. 1997b; Clough 1998; Naidoo et al. 1998, 2002; Snedaker and Araújo 1998; Theuri et al. 1999; Patanaponpaiboon and Poungparn 2000; Sobrado 2000; Mehlig 2001; Das et al. 2002; Cheeseman and Lovelock 2004; Parida et al. 2004)

Species	A	Location
Avicennia marina	22.0	Australia
Rhizophora apiculata	23.2	Malaysia
R. apiculata	20.1	Malaysia
A. marina	17.6	South Africa
Hibiscus tiliaceus	9.1	South Africa
Rhizophora mangle	11.9	Brazil
A. marina	12.5	Brazil
R. mangle	15.9	USA
Avicennia germinans	20.9	USA
Lumnitzera littorea	25.0	USA
Conocarpus erectus	22.8	USA
Bruguiera gymnorrhiza	11.8	India
R. apiculata	15.3	India
Bruguiera cylindrica	20.5	India
Ceriops tagal	3.2	Kenya
Rhizophora mucronata	4.0	Kenya
Rhizophora stylosa	7.5	Western Australia
R. mangle	6.8 (fringe)	Belize
R.mangle	5.8 (dwarf)	Belize
B. gymnorrhiza	8.3	Australia
Ceriops australis	6.1	Australia
R. apiculata	10.3	Australia
R. stylosa	12.9	Australia
Bruguiera parviflora	13.2	India
Avicennia alba	17.9	Thailand
Excoecaria agallocha	14.2	Thailand
L. littorea	17.4	Thailand
Ceriops decandra	7.0	Thailand
A. germinans	5.6	Venezuela
B. gymnorrhiza	10.6	South Africa

2.4.2 *Respiration*

There are quite a few measurements of soil respiration in mangrove forests (Section 5.4.1), but there are surprisingly few respiration measurements for mangrove leaves and for construction and maintenance respiration in roots, branches, and stems. As in other trees, we may presume that only a small percentage (≈10%) of carbon fixed is respired by non-leaf parts (Barnes et al. 1998), but data for trees from all biomes are exceedingly few (Perry et al. 2008).

Some respiration data are available for mangrove leaves and roots. The rate of dark respiration in mangrove leaves ranges from 0.2–1.4 μmol CO_2 m^{-2} s^{-1} (Table 2.6)

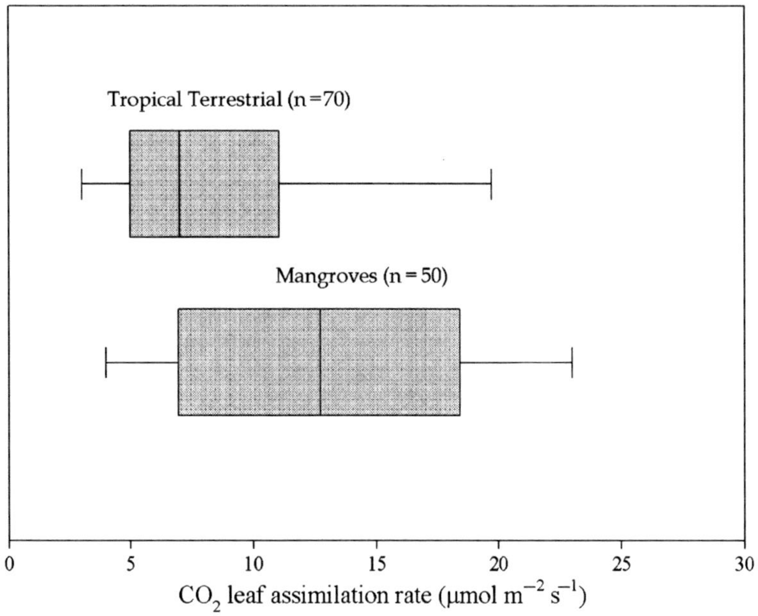

Fig. 2.10 Comparison in CO_2 leaf assimilation rates between various species of mature tropical mangrove and terrestrial trees. Mangrove data are from sources listed in Table 2.5 (plus older references within) (Terrestrial data from references cited in Fig. 2.29 in Turner 2001 plus data in Doley et al. 1987; Roy and Salager 1992; Königer et al. 1995; Krause et al. 1995; Nygren 1995; Zotz et al. 1995; Swanborough et al. 1998; Eamus et al. 1999; Ishida et al. 1999a, b; Lopez and Kursar 1999; Lovelock et al. 1999; Marenco et al. 2001; Leakey et al. 2003; Kenzo et al. 2004)

Table 2.6 Mean rates of dark respiration (R, μmol CO_2 m^{-2} s^{-1}) and photosynthesis to respiration ratio (P/R) in leaves of some mangrove species (Data from Golley et al. 1962; Lugo et al. 1975; Smith JAC et al. 1989; Gong et al. 1992; Clough 1998, unpublished data)

Species	R	P/R	Location
Rhizophora mangle	1.1	3.5	Puerto Rico
Ceriops australis	0.6	10.1	Australia
Rhizophora apiculata	1.4	7.4	Australia
Avicennia germinans	0.5	11.2	Venezuela
Conocarpus erectus	0.2 (dry season)	9.4	Venezuela
C. erectus	0.7 (wet season)	5.8	Venezuela
R. mangle	0.7	6.0	USA
Lumnitzera racemosa	1.0	3.4	USA
A. germinans	0.4	2.1	USA

with photosynthesis to respiration (P/R) ratios ranging from 2.1–11.2, which are at the upper end of the range for tropical terrestrial trees (Reich et al. 1997).

For mangrove roots, Golley et al. (1962) made the first known large-scale field measurements in *R. mangle*-dominated forests in Puerto Rico, where rates of prop root respiration averaged 169 mmol C m^{-2} prop root day^{-1}. The measurements were crude compared with recent studies, but the respiration rates for roots were the second largest loss of carbon in these Puerto Rican mangroves after leaf respiration. Scholander et al. (1955) first measured mangrove root respiration at a finer scale, but most studies have focused on the pneumatophores of *Avicennia* (Scholander et al. 1955; Burchett et al. 1984; Curran 1985; Kitaya et al. 2002). In distal and proximal roots of *A. marina*, Burchett et al. (1984) found maximal rates of root respiration (3.2 µmol CO_2 g^{-1} FW root h^{-1}) at 25% of full-strength seawater, with slower rates in freshwater (2.8 µmol CO_2 g^{-1} FW root h^{-1}) and in 50% (3.1 µmol CO_2 g^{-1} FW root h^{-1}) and 100% (2.1 µmol CO_2 g^{-1} FW root h^{-1}) seawater. In a detailed study of gas exchange and oxygen concentration in roots of *Sonneratia alba, Avicennia marina, Bruguiera gymnorrhiza*, and *Rhizophora stylosa* in Okinawa, Kitaya et al. (2002) measured net photosynthesis in the raised pneumatophores of *S. alba* and *A. marina* and in the prop roots of *R. stylosa* of 0.6, 0.2 and 0.1 µmol CO_2 m^{-2} s^{-1}, respectively. No photosynthetic activity was detected in the knee roots of *B. gymnorrhiza*. Rates of root respiration averaged 1.3, 0.8, and 2.5 µmol CO_2 m^{-2} s^{-1} in *S. alba, A. marina* and *R. stylosa*, respectively. In hydroponic culture, fine root respiration in *R. mangle* ranged from 3–6 nmol CO_2 g^{-1} s^{-1} at 25 °C (McKee 1996) and comparable rates (0.5–6 nmol CO_2 g^{-1} s^{-1}) were measured for *R. mangle* roots harvested in Belize (Lovelock et al. 2006c). The impact of N and P additions on rates of root respiration were equivocal (Lovelock et al. 2006c). These rates are low compared with estimates (3–55 nmol CO_2 or O_2 g^{-1} s^{-1}) for other angiosperm trees (Reich et al. 1998; Burton et al. 2002; Cheng et al. 2005).

2.5 Primary Productivity

Critical to our ability to estimate the role of mangroves in regional and global carbon cycling is an accurate estimation of net primary production. About 2% of the radiant energy reaching the Earth's surface is used by plants to assimilate atmospheric CO_2 into organic compounds used to construct new leaf, stem, branches, and root tissue, as well as to maintain existing tissue, create storage reserves, and to provide chemical defense (e.g., polyphenolic acids) against insects, pathogens, and herbivores. Figure 2.11 summarizes the allocation of fixed carbon by mangrove trees. Net production is the balance between gross photosynthesis and leaf dark respiration, and represents the amount of carbon available for growth and tissue maintenance (Fig. 2.12). As noted earlier, photosynthesis varies with many factors, especially light intensity, temperature, nutrient and water availability, salinity, tidal range, stand age, species composition, wave energy, and weather. Further, evidence of the long-term consequences of climate variability can be gleaned from recent findings of annual growth rings in Asian and African forests of *R. mucronata* and *R. apiculata* (Yu et al. 2004; Verheyden et al. 2004, 2005), as well as seasonal patterns of tree growth in *S. alba* and *B. gymnorrhiza* stands in Micronesia (Krause et al. 2008).

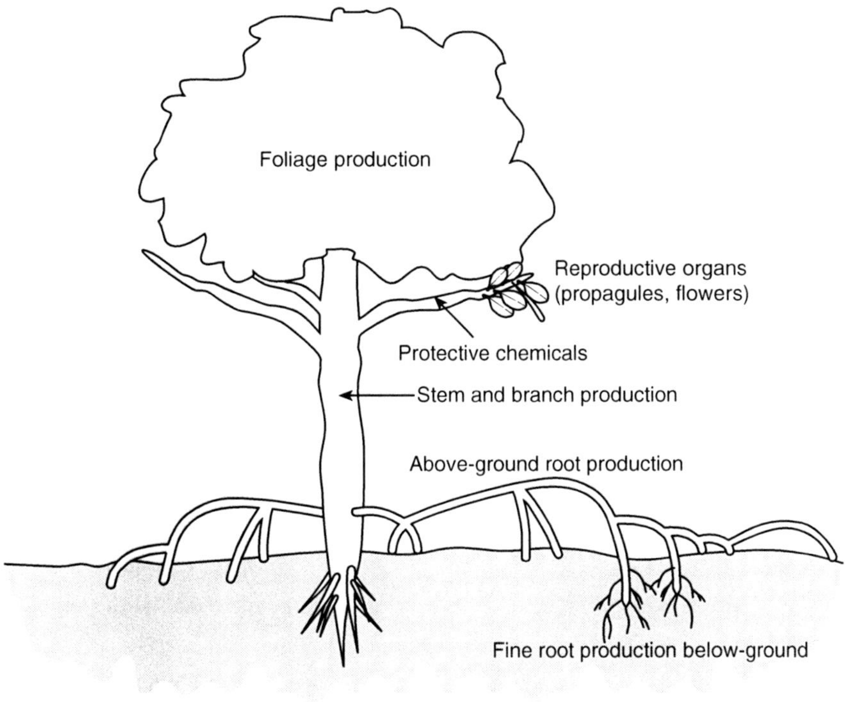

Fig. 2.11 Idealized scheme of the allocation of carbon derived from photosynthesis within a mangrove tree

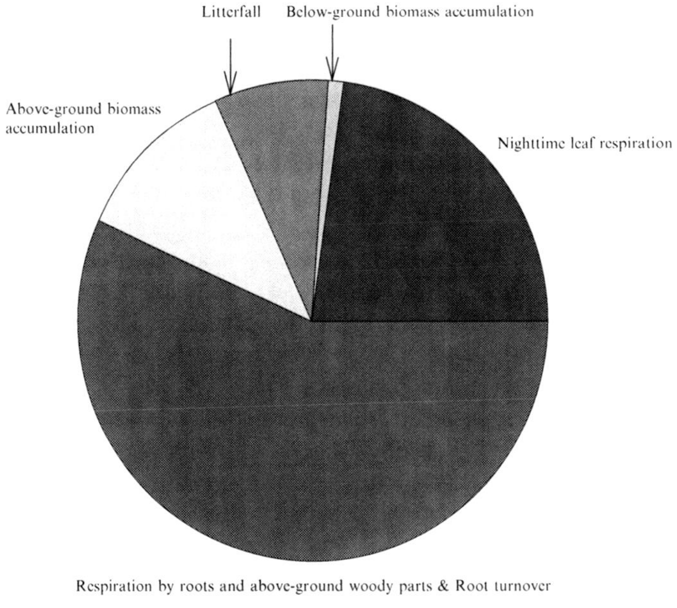

Fig. 2.12 A preliminary carbon balance for 22 year-old *Rhizophora apiculata* trees in the Matang Mangrove Forest Reserve, Malaysia (Data from Clough et al. 1997b)

2.5.1 Methods and Their Limitations

Methods used to measure terrestrial forest production have only partly been adapted for use in mangroves. This is puzzling considering that the first measurements of whole-plant CO_2 exchanges occurred early last century (Baldocchi and Amthor 2001). Traditionally, net primary production (NPP) is measured indirectly by measuring and summing (1) the biomass of the incremental growth of stems and (2) litterfall. These traditional methods make no allowance for below-ground growth and thus underestimate true NPP. More recently, models of ecosystem carbon and water flux in terrestrial forests have been fitted to eddy covariance data (see Section 6.3.1) which are measures of gas exchange from the forest floor to above the canopy. These models incorporate the Farquhar model of C_3 photosynthesis (Farquhar et al. 1980) and several empirical models of stomatal response to the environment, as well as including an estimate of heterotrophic respiration. Such whole-ecosystem methods including use of remote sensing and chlorophyll fluorescence techniques (Nichol et al. 2006), are just starting to be used for mangroves (Section 6.3.1).

Five methods most often used to measure net primary production of mangrove forests are:

- Litterfall plus incremental growth
- Harvesting
- Gas exchange
- Light attenuation/gas exchange
- Demographic/allometric changes

Litterfall is by far the most common method used because it is inexpensive and easy to measure, but it only measures leaf production and not growth of the remainder of the tree. Harvesting is labor intensive and slow, and usually available only as a result of silviculture. Like litterfall, harvesting only accounts for above-ground production; leaf production is often unaccounted for. Measurements of gas exchange are precise and rapid, but subject to the problem of extrapolating from a small area—usually a few individual trees—to an entire stand, introducing the problem of error. Moreover, relying solely on gas exchange measurements overestimates net production as it does not account for most tree respiration. Combining measurements offers the best hope of accounting for production of all, or most, tree parts. Litterfall and incremental growth of the trunk account for nearly all above-ground production (but not below-ground production).

Arguably one of the best methods to measure primary production is to measure light attenuation through the canopy. The early efforts (e.g., Bunt et al. 1979) provided rapid and relatively easy estimates of potential net primary production, but suffered from lack of actual photosynthesis measurements and relying on a number of untested assumptions based on light attenuation models from temperate forests (Kirita and Hozumi 1973). The method relies on relating the amount of light absorbed by the mangrove canopy to the total canopy chlorophyll content. Three workers (J. Ong, B. Clough, and W. Gong) subsequently modified the light attenuation method, combining measurement of light attenuation with a more robust method of calculation of photon flux density at the bottom of the canopy and empirical measurements of leaf photosynthesis (Gong et al. 1991, 1992; Clough 1997; Clough et al. 1997b). This modified method still relies on measurements of light absorption by the forest canopy to estimate leaf area index, which is the amount of leaf area relative to the amount of ground area. Leaf area index is calculated using the formula:

$$L = [\log_e (I)_{mean}] - [\log_e (I_o)_{mean}]/-k$$

where $(I)_{mean}$ = the mean photosynthetically active radiation (PAR) under the canopy, $(I_o)_{mean}$ = incident PAR, and k = canopy light extinction coefficient. A large number of measurements by Clough (1997) found that k commonly lies between 0.4 and 0.65 in a variety of mangrove canopies, with an average of ≈0.5. The leaf area index (L) is then used to estimate net daytime canopy photosynthesis (P_N) using the formula:

$$P_N = A \times d \times L$$

where d = daylength (h) and A = average rate of photosynthesis per unit leaf area, which is obtained by measurement of CO_2 exchange (as in Table 2.5). Using this modified method, Clough et al. (1997b) compared their more robust estimates with the earlier light attenuation method (Table 2.7). The comparison shows that the original method underestimates the net canopy production by a factor of 12. This suggests that many published measurements using the Bunt et al. (1979) method need to be corrected by this factor, clearly increasing previous estimates of mangrove canopy production.

Comparing the gas exchange, litterfall, increment growth plus litter fall, the original light attenuation, and the modified light attenuation methods illustrates the differences among methods (Table 2.8), and the difficulty in our ability to settle on an accurate range of net primary production values for mangrove forests.

It is clear that litterfall underestimates and net gas exchange overestimates, net primary production. From the remaining data, the modified light attenuation method appears to give a reasonable estimate of total production while litterfall

Table 2.7 Comparison of net canopy production (t DW ha^{-1} $year^{-1}$) derived from the method described in Bunt et al. (1979) and the modified procedure described in Clough (1997) and Clough et al. (1997b). Measurements were made in a 22-year old *Rhizophora apiculata* forest in peninsular Malaysia

Measurement set	Light attenuation method	Modified procedure
1	11.0	135
2	13.0	161
3	13.7	165
4	14.3	157
Mean	13.0 ± 1.4	155 ± 13

Table 2.8 Estimates of net primary production (t DW ha^{-1} $year^{-1}$) of *R. apiculata* forests of various age in peninsular Malaysia using five of the most used procedures (Data from Gong et al. 1984, 1992; Ong et al. 1995; Clough et al. 1997b; Alongi et al. 2004a)

Age	Net gas exchange	Litter fall	Litterfall + incremental growth	Light attenuation	Modified light attenuation
5	132	7	19	14	37
10	122	10	34	19	
20	240	10	30	16	65
70	NA	8	NA	21	102

plus incremental growth yield estimates of above-ground production (but excluding below-ground production) comparable to those measured in terrestrial forests. The modified light attenuation method measures total net fixed carbon production in the daytime and offers the most robust assumptions based on tree physiology and carbon balance, but it may not provide a reliable estimate of net primary production because: (1) the relationship between potential net canopy production and the actual net primary production has yet to be determined, (2) the average rate of photosynthesis per unit leaf area should be measured at each site, and (3) the method does not account for respiration by the branches, stem or roots, or dark respiration by the leaves. Values obtained by this method can best be described as net daytime canopy photosynthesis.

A number of recent studies have attempted to measure above-ground production using allometry coupled with litterfall or leaf turnover (Duarte et al. 1999; Coulter et al. 2001; Ross et al. 2001; Sherman et al. 2003). The method employed by Ross et al. (2001) is an adaptation of methods used for grasslands, incorporating detailed allometric measurements of individual trees coupled with observations of leaf demography to measure leaf turnover. Although they were not able to compare their method directly with other procedures, their net production values are at the upper end of the range for similarly-sized forests. Coulter et al. (2001) similarly employed analysis of leaf nodes to produce an estimate of new leaf production. Combining leaf production with estimates of the number of inflorescence scars produced by the shedding of leaves and reproductive structures, above-ground production of *Kandelia candel* in Vietnam was estimated at rates comparable to, or greater than, previous values (Coulter et al. 2001).

2.5.2 Carbon Allocation of Primary Productivity

Empirical estimates of root production are few (McKee and Faulkner 2000; Gleason and Ewel 2002; Cahoon et al. 2003; Sánchez 2005). In mangrove forests of Micronesia, Gleason and Ewel (2002) measured growth of fine roots into chambers placed in holes left by coring for root biomass. Over the upper 30 cm of soil, they measured growth rates of 0.29, 0.33 and 2.78 mg DW cm^{-3} $year^{-1}$ in *R. apiculata, B. gymnorrhiza*, and *S. alba* forests. The other studies also utilized the in-growth core technique (McKee and Faulkner 2000; Cahoon et al. 2003; Sánchez 2005) measuring rates of root production ranging from 18–1,145 g DW m^{-2} $year^{-1}$, but with most values between 307–378 g DW m^{-2} $year^{-1}$. These estimates are at the lower end of the range compared with similar measurements in tropical terrestrial forests (Clark et al. 2001; Perry et al. 2008), but most of these measurements were made in fringe stands. It is likely that mangrove roots grow more than 1–2 m into the soil (McKee et al. 2007) and more rapidly in more luxuriant forests. An analysis of soil respiration over 10° of latitude suggests that mangroves allocate proportionally more carbon below-ground than terrestrial trees (Lovelock 2008). As we will investigate in Section 7.1.1, below-ground production is likely to be much higher than previously believed.

Wood production, in contrast, is well known, with a plethora of data available from harvesting studies. Saenger (2002) provides a complete analysis of the mean annual growth of mangrove wood. Regressing stand age with increases in above-ground biomass, he found a significant positive regression coefficient of 0.65. The data indicate that mangrove trees in plantations where the climate is favorable can grow rapidly compared to trees in terrestrial plantations (see chapters in Nambiar and Brown 1997). Regressing mean growth rates of mostly *Rhizophora apiculata* trees in variously aged plantation stands throughout southeast Asia, Saenger (2002) found a more complex relationship with age, with peak growth at about 15 years, declining thereafter. The relationship is described as $y = -0.041x^2 + 1.342x + 1.101$, where y = mean annual increment (t ha^{-1} $year^{-1}$) and x = stand age (year). Across all species and ages, mean annual increment of mangrove trees ranges from 0.1– 1.8 cm $year^{-1}$ (at diameter-at-breast height).

The estimates of net canopy production made using the modified light attenuation method may include below-ground production (this is unclear at present) but there is no thorough perception as to how carbon is allocated within a mangrove tree. Clough et al. (1997b) provides a preliminary carbon allocation model for 22 year-old *Rhizophora apiculata* trees in Malaysia (Fig. 2.12). Of a total annual net daytime fixed C production of 56 t C ha^{-1} $year^{-1}$, 22% is respired by the foliage overnight, 11% is accumulated as above-ground biomass, 8% is lost as litter, 1% is accumulated as below-ground biomass, and, by difference, the remaining 58% is presumably used in root turnover, and in respiration of branches, stem, roots, and other woody parts.

As pointed out by Clough (1998), it is not yet possible to construct a robust model of carbon balance for mangrove trees owing to the lack of empirical data and the difficulty of measuring root processes and respiration of woody parts. However, their preliminary carbon allocation budget suggests that roughly half of assimilated carbon is eventually respired, in agreement with similar estimates for terrestrial trees (Barnes et al. 1998; Clark et al. 2001; Perry et al. 2008).

At the canopy level, some estimates exist of the balance between net assimilation and respiration. The early data from Florida mangroves (summarized in Odum et al. 1982) show that canopy respiration accounts for an average of 58% of gross primary production, with a range of 14.3–85.9%. In Malaysian forests, Alongi et al. (2004a) estimated that tree respiration equates to 41% of gross primary productivity. Canopy photosynthesis and respiration have rarely been measured simultaneously (Clough 1992). On Okinawa, Suwa et al. (2006) made extensive measurements of gross photosynthesis and respiration throughout the canopy of a *Kandelia candel* forest. They found maximum rates of both GPP and dark respiration at the top of the canopy with a two- to sevenfold decline to the bottom of the foliage. Annual canopy gross primary production averaged 102.9 t CO_2 ha^{-1} $year^{-1}$ and canopy respiration averaged 44 t CO_2 ha^{-1} $year^{1}$, for an average annual net primary production of 58.9 t CO_2 ha^{-1} $year^{-1}$ or 57% of GPP. The similarity among results is most likely a reflection of physiological limits of carbon assimilation and allocation as well as constraints imposed on physiological processes by the laws of thermodynamics.

2.5.3 Rates and Patterns of Net Primary Productivity

Although caution must be applied when considering net primary productivity estimates based on a variety of methods used in disparate settings in forests of different age and living under different environmental conditions, the available data (Table 2.9) suggests that rates of net primary production of mangrove forests are rapid compared with other estuarine and marine primary producers (Gattuso et al. 1998; Duarte et al. 2005).

If we accept the data obtained using the modified light attenuation method as a reliable estimate of net primary productivity of mangroves in the daytime, the average rate of net primary production averages 64 t DW ha^{-1} $year^{-1}$. In comparison, the estimates based on incremental growth plus litterfall averages 11 t DW ha^{-1} $year^{-1}$. The former figures do not represent the true rates of net primary production, but they do suggest that mangroves are more significant carbon fixers than previously thought, at least in the daytime. The problem is that it is still unclear what exactly is being measured using the light attenuation method in mangrove forests. Unlike the above differences in estimates for mangroves, modelling of light interception in terrestrial forests results in NPP estimates only slightly higher than those obtained using more traditional harvesting methods (Grace et al. 2001). Nevertheless, plotting the data from the light attenuation method versus latitude (Fig. 2.13) gives a significant negative relationship, indicating that mangrove productivity declines away from the equator, mirroring the latitudinal decline in mangrove biomass (Fig. 2.6) and litterfall (Saenger and Snedaker 1993).

Contrary evidence to these patterns has been provided by Lovelock et al. (2007) based on data of plant growth and nutrient content in leaves in fertilized plots between latitudes 36° S and 27° N. Two hypotheses to explain increases in plant N and P concentrations with latitude were tested: (1) the geochemical hypothesis, which proposes that geochemical limitation to P availability in the tropics is the cause, and (2) the growth rate hypothesis, which states that greater growth rates

Table 2.9 Estimates of net primary production (NPP = t DW ha^{-1} $year^{-1}$) of mangrove forests in various parts of the world based on different methods (Data from Golley et al. 1962; Miller 1972; Hicks and Burns 1975; Lugo et al. 1975; Christensen 1978; Ong et al. 1984, 1985, 1995; Twilley 1985a, b; Putz and Chan 1986; Aksornkoae et al. 1989; Lee 1990; Atmadja and Soerojo 1991; Gong et al. 1991, 1992; Robertson et al. 1991; Amarasinghe and Balasubramaniam 1992; Sukardjo and Yamada 1992; Sukardjo 1995; Day et al. 1996; Clough et al. 1997b,1999; Clough 1998; Cox and Allen 1999; Alongi and Dixon 2000; Alongi et al. 2000a, 2004a; Kathiresan 2000; Coulter et al. 2001; Ross et al. 2001; Sherman et al. 2003; Hossain et al. 2008)

Species	Location	NPP	Method
R. mangle, A. germinans, L. racemosa	USA	46.0	Gas exchange
R. mangle, A.germinans, L. racemosa	USA	26.1 (fringe) 8.1 (dwarf)	Demographic/ allometric
A. germinans	USA	20.5	Gas exchange
R. mangle	USA	16.9	Gas exchange
R.mangle, A. germinans, L. racemosa	USA	22.5	Gas exchange
R.mangle, A. germinans, L. racemosa	Puerto Rico	58.4	Gas exchange
R. apiculata	Thailand	63.7[a] (13.1)	Light attenuation
C. decandra	Thailand	48.7[a] (9.7)	Light attenuation
R. apiculata	Malaysia	112.1[a]	Light attenuation
R. apiculata (70 years)	Malaysia	102.2[a] (24.6)	Light attenuation
R. apiculata (18 years)	Malaysia	65.7[a] (14.7)	Light attenuation
R. apiculata (5 years)	Malaysia	36.5[a] (12.8)	Light attenuation
B. parviflora	Malaysia	27.4	Harvest/incremental growth
R. mangle (5 years)	Cuba	1.6[b]	Harvest/incremental growth
A. germinans	Cuba	5.9[b]	Harvest/incremental growth
L. racemosa	Cuba	5.4[b]	Harvest/incremental growth
Sonneratia apetala	Bangladesh	12.5[b]	Harvest/incremental growth
Sonneratia caseolaris	Bangladesh	26.4[b]	Harvest/incremental growth
Avicennia officinalis	Bangladesh	7.6[b]	Harvest/incremental growth
A. marina	Bangladesh	4.4[b]	Harvest/incremental growth
A. alba	Bangladesh	2.1[b]	Harvest/incremental growth
B. gymnorrhiza	Bangladesh	0.6[b]	Harvest/incremental growth
Bruguiera sexangula	Bangladesh	0.1[b]	Harvest/incremental growth
Excoecaria agallocha	Bangladesh	4.7[b]	Harvest/incremental growth
Xylocarpus moluccensis	Bangladesh	0.5[b]	Harvest/incremental growth
Mixed species	Micronesia	4.2[b]	Harvest/incremental growth
R. apiculata, B. gymnorrhiza	Malaysia	8.7[b]	Harvest/incremental growth
R. apiculata	Vietnam	4.9[b]	Harvest/incremental growth
R. apiculata	Vietnam	19.0	Incremental growth
R. apiculata	Thailand	15.7	Incremental growth
R. apiculata	Thailand	10.6	Incremental growth
R. apiculata	Vietnam	9.4	Litterfall
R. apiculata	Vietnam	18.7	Litterfall
R. racemosa	Gambia	18.8	Litterfall
Avicennia africana	Gambia	11.6	Litterfall

(continued)

Table 2.9 (continued)

Species	Location	NPP	Method
R. racemosa	Gambia	10.4	Litterfall
R. mucronata	India	14.6	Litterfall
R. apiculata	India	13.6	Litterfall
A. marina	India	6.2	Litterfall
B. sexangula	China	11.0	Litterfall
Kandelia candel	China	13.3	Litterfall
K. candel	China	24.4	Litterfall/allometric
R. mucronata	Indonesia	23.4	Litterfall/incremental growth
R. apiculata	Thailand	13.5	Light attenuation
Aegiceras corniculatum	China	11.3	Litterfall
K. candel	Vietnam	5.3	Demographic/allometric
K. candel	Vietnam	13.4	Demographic/allometric
R. stylosa	Australia	40.5[a] (9.6)	Light attenuation
A. marina	Australia	30.6[a] (6.4)	Light attenuation
Mixed *R. mangle, A. germinans, L. racemosa*	Dominican Rep.	19.7[c]	Demographic/allometric
Mixed *R. mangle, A. germinans, L. racemosa*	Guadeloupe	21.2 (fringe) 6.2 (dwarf)	Litterfall/incremental growth
R. mangle	Hawaii	29.1	Litterfall/incremental growth
Mixed *Rhizophora* spp.	Australia	29.2	Light attentuation
R. mucronata/A. marina	Sri Lanka	11.0	Litterfall/incremental growth
R. apiculata, B. parviflora	Papua New Guinea	30.5[a] (9.7)	Light attenuation
Nypa fruticans	Papua New Guinea	30.1[a] (9.9)	Light attenuation
A. marina, Sonneratia lanceolata	Papua New Guinea	24.4[a] (6.8)	Light attenuation
R. apiculata, A. marina	Indonesia	104.6	Light attenuation
R. apiculata, A. marina	Indonesia	96.9	Light attenuation
A. officinalis, A. marina	Indonesia	103.2	Light attenuation
C. tagal, R. apiculata	Indonesia	106.1	Light attenuation
C. tagal, R. apiculata	Indonesia	109.4	Light attenuation
R. stylosa, S. alba	Indonesia	63.7	Light attenuation
R. apiculata, K. candel	Indonesia	74.3	Light attenuation

[a]Estimate using the modified light interception method or original data recalculated using the modified method (see text). A correction factor of 4.8 was applied based on the data in Table 2.7 and all data calculated with both light inception methods (all those in above table asterisked[a]). Of n = 11 forests, the original method gave a mean NPP estimate of 11.85 t DW ha^{-1}year^{-1} and the modified method gave a mean NPP value of 57.08 t DW ha^{-1} year^{-1}, for an average difference of 4.8. All C values were converted to DW assuming that mangrove wood is 48% C by DW (Alongi et al. 2003a). Estimates based on the original light interception method are in parentheses.
[b]Assumes an average density of 0.9332 t m^{3} (Saenger 2002).
[c]Sherman et al. (2003).

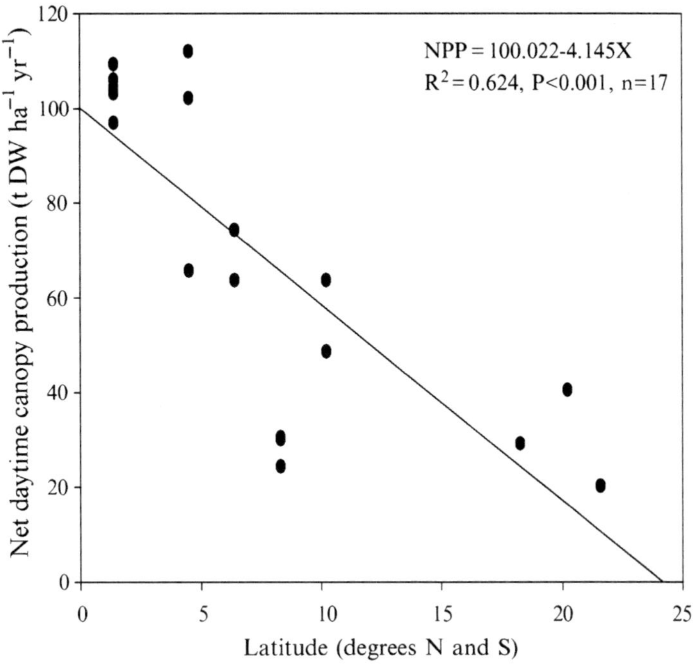

Fig. 2.13 Latitudinal changes in net daytime canopy production measured using the modified light interception method (Data from Atmadja and Soerojo 1991; Gong et al. 1991, 1992; Robertson et al. 1991; Sukardjo 1995; Clough et al. 1997b; Clough 1998; Alongi and Dixon 2000; Alongi et al. 2000a, c, 2004a)

requiring high nutrient levels are needed to complete growth and reproduction during shorter growing seasons in temperate regions than in the tropics. Lovelock et al. (2007) shows that temperature-adjusted growth rates of trees significantly increase with latitude, supporting the growth rate hypothesis. However, both nutrient resorption efficiency and photosynthetic P-use efficiency decreased with increasing latitude, suggesting less P limitation at higher latitudes, offering some support also for the geochemical hypothesis. Both hypotheses may not be mutually exclusive, considering that low P availability in the tropics has likely been a key evolutionary driver in selecting for plant traits relating to nutrient-use efficiency (see Section 2.5.4).

How do these productivity data compare with productivity data for tropical rain forests? First, we must compare data obtained using identical or very similar methods. The most comprehensive database for both mangroves and tropical terrestrial forests involves measurement of above-ground biomass accumulation plus litterfall. Comparing the data in Table 2.9 and the data analyzed by Clark et al. (2001) and Scurlock and Olson (2002), we find equivalent estimates (Fig. 2.14). For mangroves (n = 29), the mean rate of above-ground net primary production is 11.13 t DW ha^{-1} year^{-1} (=44.52 mol C m^{-2} year^{-1} assuming 48% C content of dry wood, Alongi et al.

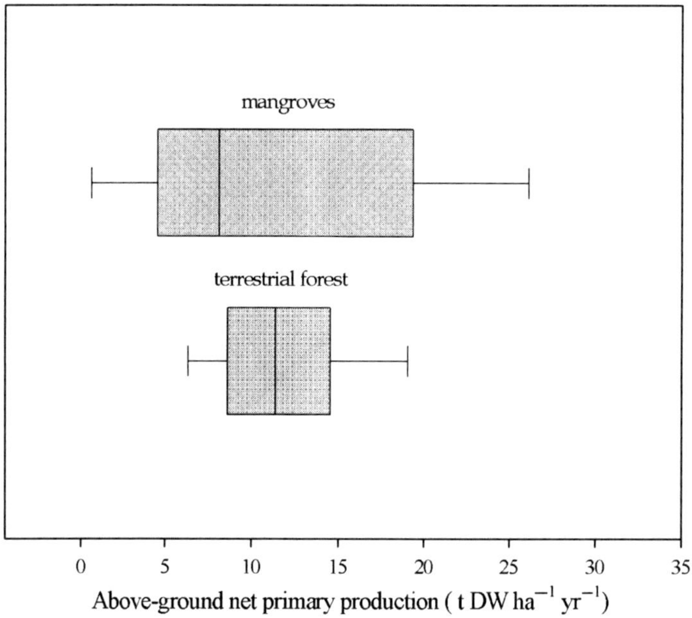

Fig. 2.14 Comparison of above-ground net primary production in mangrove and tropical terrestrial forests based on measurements of biomass increments and litterfall. Vertical line in box denotes median and the boxes encompass the 25th and 75th percentiles and the outer bars denote the 5th and 95% percentiles, respectively (Data from Table 2.9 for mangroves and from Clark et al. 2001 and Scurlock and Olson 2002 for terrestrial forests)

2003a) with a median value of 8.1 and 25th and 75th percentiles of 4.6 and 19.175, respectively. For terrestrial forests, the mean rate of above-ground NPP is 11.9 t DW ha^{-1} year^{-1}; the median value is 11.4, the 25th percentile is 8.8 and the 75th percentile is 14.4. Given the differences within and between both forest groups in size, age, and species differences, the values are remarkably close, suggesting that rates of primary production are equivalent between mangrove and tropical terrestrial forests. It also underscores the similarities in physiological and ecological factors limiting production of all trees, although one must remember that rates of below-ground production are sorely lacking for all forests (Komiyama et al. 2008). Respiration of roots and woody parts are also badly needed to adjust the photosynthetic rates for a true estimate of net carbon fixation (Komiyama et al. 2008).

Like other forests, mangrove stands vary in size and age over time, and therefore vary in rates of production and in the balance between production and respiration. Long-term patterns are important to discern as they reflect a balance between factors promoting and limiting forest growth. A few studies have examined the growth dynamics of mangrove forests over time or of stands of known age (Ong et al. 1985; Day et al. 1996; Fromard et al. 1998; Clough et al. 2000; Alongi 2002). In a mangrove forest bordering Laguna de Terminos in Mexico, Day et al. (1996) measured litterfall plus annual biomass increments in: (1) zones of *A. germinans* with *R.*

mangle as a canopy sub-dominant, (2) a scrub forest of *A. germinans*, and (3) a stand of mature *A. germinans*. Interannual patterns of litterfall correlated best with patterns in soil salinity, precipitation, and air temperature, with these three factors explaining 74% of the variance. Interannual patterns of total above-ground production did not correlate significantly with climatic variables, suggesting that seasonal and annual changes in solar radiation, temperature, rainfall, evapotranspiration, daylength, or other factors, such as time lags, all play roughly equal roles in affecting net primary production.

It is often overlooked that mangroves, like other forests, change over time through successive stages of development with sequential changes in species and therefore, in rates of net primary production. A good example of how mangrove forests change over time comes from the study of mangroves along the coast of French Guiana, which is greatly influenced by the Amazon (Fig. 2.15). The data are expressed in terms of stem density but rates of above-ground production likely mirror the changes in tree density, with a rapid increase in the growth and density of pioneering species over the first 5 years, followed by maturation over about 5 decades with a clear decline after about 70 years (Fromard et al. 1998). This scheme mirrors an earlier one proposed by Jiménez et al. (1985) based on patterns of dead trees in mangrove forests worldwide.

The stable-state maturity phase of the French Guiana forests appears to be prolonged compared with what is known for most terrestrial forests (Barnes et al. 1998; Perry et al. 2008). The long maturity phase may represent an alternate succession state in which the time clock for the climax stage of the forest is "reset" by each major disturbance. Indeed, the relationship between forest age and photosynthetic production in mangroves suggests that this prolongation, or arrested progression, is

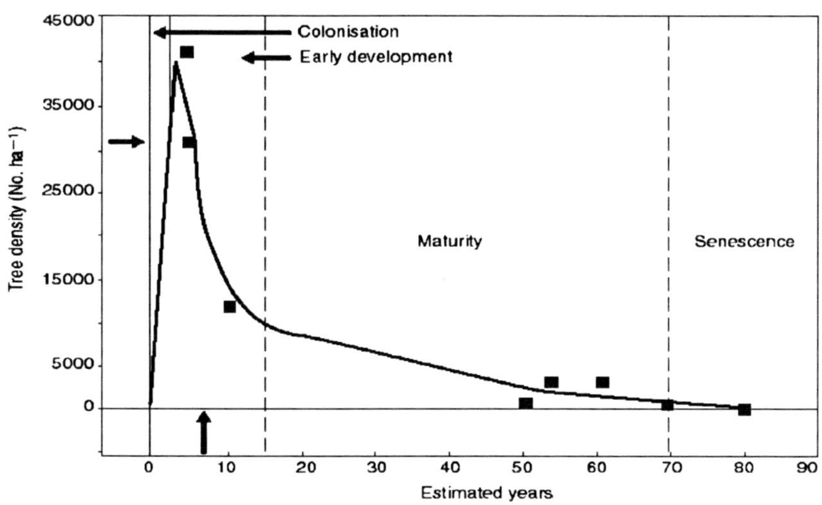

Fig. 2.15 Evolution of a mangrove forest over time (Data and model modified from Fromard et al. 1998)

what happens when forests are disturbed (Alongi 2002, 2008; Piou et al. 2008). A plot of net canopy production versus age of various forests of *Rhizophora apiculata* (Fig. 2.16) in Southeast Asia shows log-phase production until about 20 years, after which NPP levels off but does not noticeably diminish for nearly a century. All of these data were obtained from sites where forests are harvested or disturbed in some way (herbicides, etc); even the oldest site was thinned 85 years ago. Other data for *Rhizophora apiculata* plantations (Clough et al. 2000) indicate a significant decline in leaf area index in stands ranging in age from 6 to 36 years, although the drop was relatively small (from 4.9 to 3.3). Few forests are pristine in the tropics, but mangroves might constitute a carbon sink for up to a century if left undisturbed.

2.5.4 *Nutrient Limitation and Nutrient-Use Efficiency*

The growth and production of all plants, depending on habitat and environmental circumstance, is limited by micro- and macronutrients. All trees, including mangroves, have a requirement for minerals in order to synthesize cell contents to manufacture structural and reproductive tissue (Aerts and Chapin 2000).

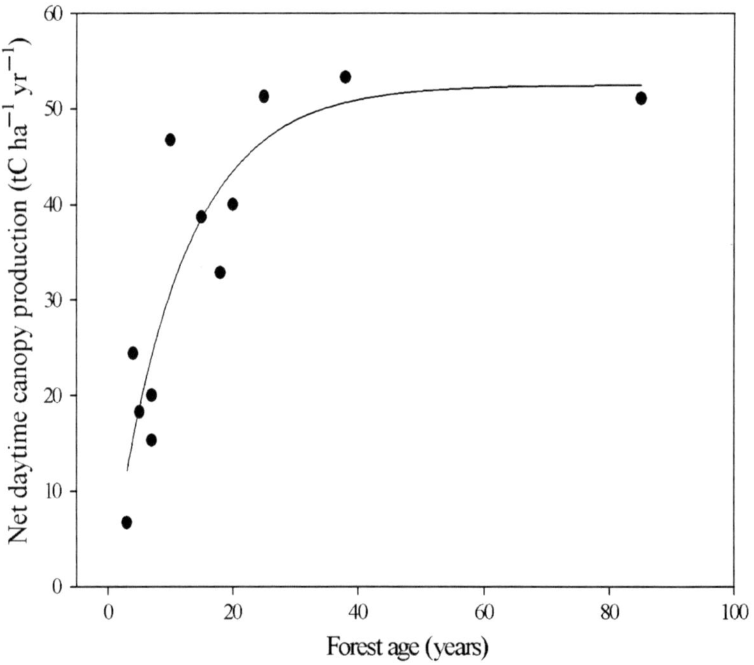

Fig. 2.16 The relationship between forest age and net daytime canopy production of *Rhizophora apiculata* in Thailand, Malaysia, and Vietnam (Data from Clough et al. 1997b, 1999; Alongi and Dixon 1999; Alongi et al. 2004a)

The micro-nutrients required are: iron, manganese, copper, zinc, nickel, molybdenum, boron, chloride, sodium, silicon, cobalt, selenium, and aluminum. The macro-nutrients are: nitrogen, phosphorus, sulfur, magnesium, calcium, and potassium. Owing to their estuarine and marine existence, mangroves are rarely limited by the relatively large quantities of sulfur, boron, potassium, magnesium, and sodium in seawater (Ball et al. 1987; Ball 1988; Boto 1991).

The critical need for nitrogen has been repeatedly demonstrated in laboratory culture for a variety of mangrove species (Clough et al. 1983; Boto et al. 1985; Naidoo 1987, 1990; Saberi 1992; Hwang and Chen 2001; Yates et al. 2002). In reality, interactive effects between different nutrients and environmental factors, such as salinity, play a key role in plant nutrition and nutrient availability. In recent laboratory factorial experiments, Yates et al. (2002) grew three mangrove species (*A. marina, C. tagal, R. stylosa*) at three levels each of nitrogen, potassium, and phosphorus and two levels of salinity. Their results showed that increasing N resulted in increased leaf numbers in all three species; interactive effects between N, P, and K were apparent, with the response differing between species. For instance, enhancement of leaf expansion rates for *C. tagal* by added N was clear only at low P concentration. This result was contrary to the results for *Kandelia candel* (Hwang and Chen 2001) wherein the addition of phosphorus stimulated growth only when N was also present. A criticism often leveled at laboratory culture studies is that nutrients are supplied at concentrations that are not limiting (or only marginally limiting) to seedlings growing in otherwise near-optimal conditions. Also, the nutritional requirements of seedlings are likely to be different than for saplings and older trees (Aerts and Chapin 2000). However, these studies are important in that they underscore the basic complexity of plant-nutrient relationships.

In the field, mangroves have demonstrated a variety of responses to added nitrogen and phosphorus, differing in relation to soil type and texture, salinity, frequency of tidal inundation, and species composition (Boto 1991). In one of the early direct tests to determine whether N and/or P are limiting to mangroves, Boto and Wellington (1983) added N or P to mixed *Rhizophora* forests in northern Australia for 1 year. They found that while N limitation occurred across the intertidal, P limitation was problematical, being apparent only in the high-intertidal forest where soils contained low levels of P.

The level of complexity of the nutrient limitation issue can be discerned from the extensive field studies conducted by Koch and Snedaker (1997) and by Feller and her colleagues (Feller 1995; McKee et al. 2002; Feller et al. 1999, 2002, 2003, 2007; Lovelock and Feller 2003; Lovelock et al., 2006a–c). In Belizean mangroves in oceanic settings with minimal terrestrial input, forests are limited by comparatively low concentrations of P, and to a lesser extent, of N, although there appears to be a switch from N to P limitation on small islands from the shore to the interior of each island. The most consistent pattern for Caribbean mangroves is that *Rhizophora mangle* is N-limited seawards, dwarf trees in the interior are P-limited, and trees across the transition from low-to high-intertidal are co-limited. Trees under P limitation are water deficient, showing more pronounced changes in structure and function when P deficiency is eased than those trees under a regime of N limitation (Lovelock et al. 2006a–c). In Florida, however, mangroves are limited only by N across the intertidal seascape (Feller et al. 2003). The data for these fertilization studies demonstrate either N- or P-limitation, or both, depending upon species composition, extent of terrigenous input, soil fertility and texture, soil redox status, and salinity, to name but a few factors regulating mangrove-nutrient relations. Recent studies have also emphasized the importance of flooding and draining of tidal waters on mangrove growth and nutrient use as tidal changes relate to rates of sediment and nutrient input (Krauss et al. 2006, 2007a, b).

The generally high photosynthetic rates for mangroves are supported by a high requirement for nutrients, implying high nutrient-use efficiencies and high rates of leaf resorption. In comparing the available mangrove data with those compiled by Aerts and Chapin (2000) and for fast-growing tropical plantation trees by Hiremath et al. (2002), it appears that while there is a wide range of values of nutrient-use efficiency for nitrogen and phosphorus for all forest types (Fig. 2.17), rates of

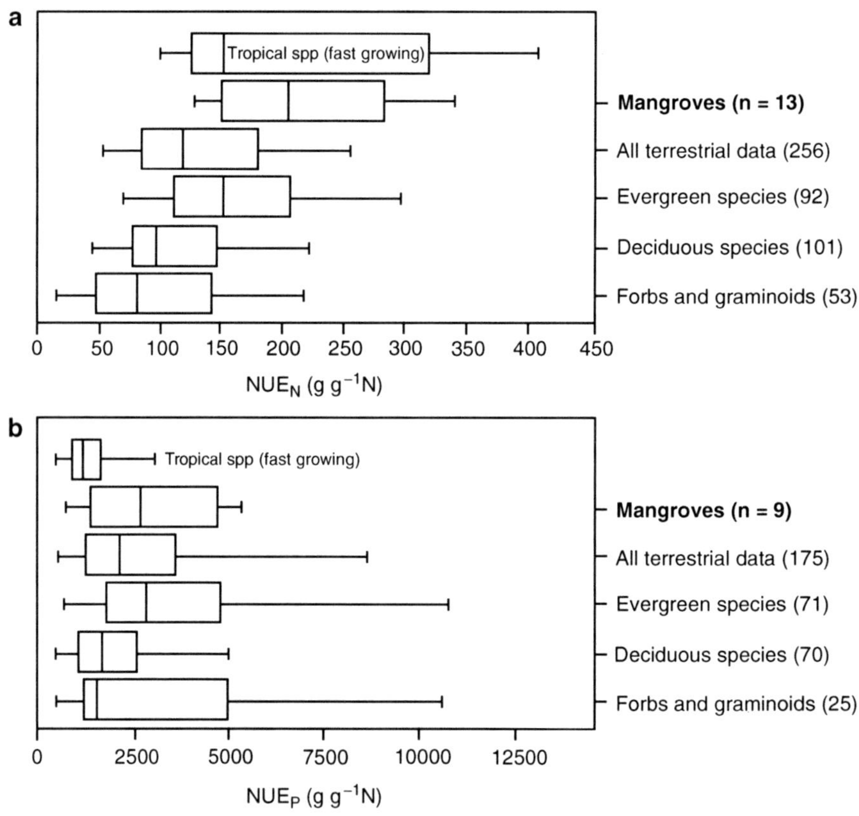

Fig. 2.17 Comparison of rates of nutrient-use efficiency in mangroves, tropical plantation trees, and all other forest types. Values in parentheses denote number of locations (Modified from Aerts and Chapin 2000. With plantation data from Hiremath et al. 2002 and mangrove data from Lugo et al. 1988; Feller et al. 1999, 2003; Lovelock et al. 1999, 2006b; Lovelock and Feller 2003; Alongi et al. 2005b)

nitrogen-use efficiency for mangroves are at the upper end of the range. Rates of phosphorus-use efficiency in mangroves however are well within the range of values for all forests. Why? Perhaps the best answer lies in the generally rapid rates of photosynthesis for mangroves which inhabit environments that, at first glance, would be inimical to fast growth. Living in oligotrophic, anoxic soils and waters, and continually inundated by tides, mangroves must be efficient to survive.

In such habitats, mechanisms to conserve limiting nutrients are clearly advantageous. The apparent strategy of using nutrients efficiently relates well to the generally low concentrations of nutrients in mangrove leaves and other tree parts (Section 2.2), and the generally high efficiency with which nitrogen and phosphorus are resorbed from leaves (Fig. 2.18). In comparing the mangrove data with those from other forested ecosystems, nitrogen is resorbed by mangrove trees at a level of efficiency at the higher end of the range. Efficiency of P resorption is within the

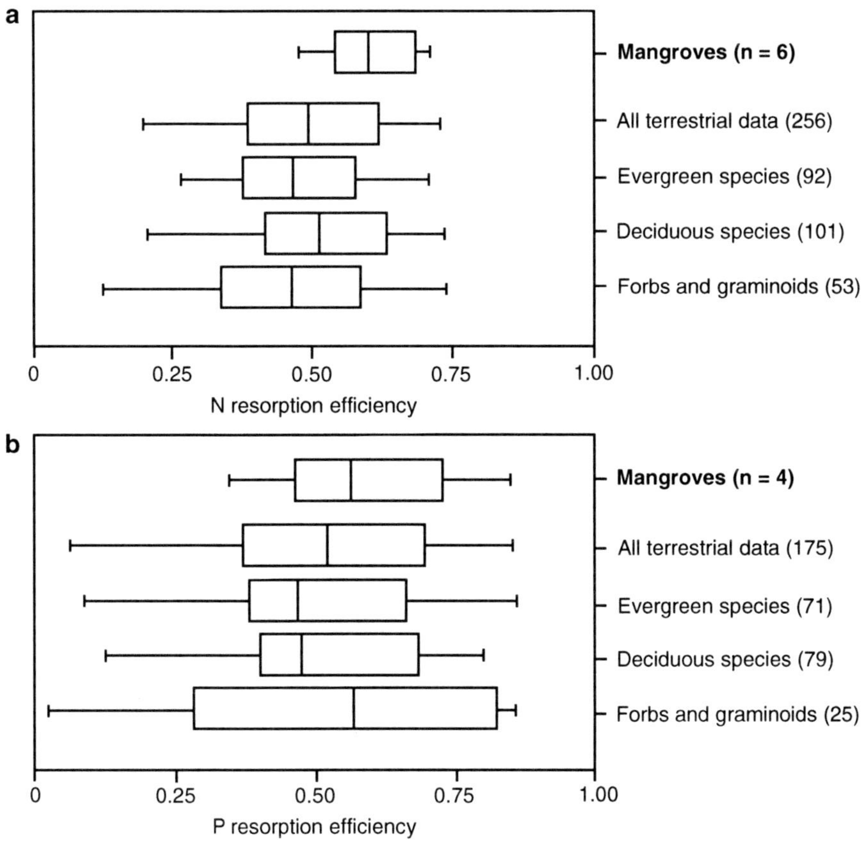

Fig. 2.18 Comparison of resorption efficiency from leaves of nitrogen (**a**) and phosphorus (**b**) in mangroves and other forested ecosystems (Data were taken from same sources listed in Fig. 2.17 caption)

mid range of other forests. As noted earlier, mangroves have other mechanisms to retain nitrogen, including a large reservoir of dead roots below-ground. Half of these mangrove values are derived from arid-zone forests, and as these trees show differences in the allocation of biomass, recycling mechanisms are probably highly developed in these forests. As in other forested ecosystems, the high rates of nutrient-use efficiency are reflected in high rates of nutrient productivity and in shorter residence times of the N and P pools (Alongi et al. 2005a).

There is also some evidence of species differences in nutrient-use efficiency (Yates et al. 2002; Lu et al. 2004; Lovelock and Feller 2003). In Western Australia, Alongi et al. (2005a) measured mean residence times for both N and P in *R. stylosa* forests of 2–4 years, whereas the residence times were shorter (<2 year) in *A. marina* stands. These differences may reflect the fact that *R. stylosa* uses a nutrient retention strategy which translates into greater rates of net primary production. There may be other reasons to explain the species differences in nutrient use:

- Differences in the way species allocate nutrients.
- Differences in the proportion of energy and nutrients vested in chemical defenses.
- Differences in leaf life spans.
- Differences in soil physicochemistry and biogeochemistry.

2.5.5 Other Primary Producers

Chlorophytes, diatoms, phytoflagellates, and cyanobacteria living on the surfaces of soils, and as epiphytes on leaves, decomposing wood, and on living roots, as well as macroalgae living mostly on prop roots, are alternative sources of fixed carbon in mangrove forests. Under well-developed canopies, algal production is dwarfed by tree production due to severe light limitation (Alongi 1994). There is some evidence that algae are negatively affected by soluble tannins leached from mangrove soils and decaying tree parts (Cooksey and Cooksey 1978). In open canopies, and especially in polluted systems, algal production is proportionally greater due to more light and greater nutrient availability. The comparatively small contribution of fixed carbon from algae belies their trophic importance as most consumers prefer algal to detrital foods (see Sections 4.6 and 5.3).

Cyanobacteria exhibit sharp horizontal zonation on prop roots and pneumatophores, and are capable of high rates of nitrogen fixation (Potts 1979; Potts and Whitton 1980). A unique nitrogen-fixing flora of Rivilariaceae is found on the pneumatophores of *Avicennia marina* located in the Red Sea, with an entirely different flora of non-heterocystous algal types on the soil surface (Potts 1979). In the Red Sea mangroves where shrub canopies are open, cyanobacteria are abundant (5.9–32.7 μg chl *a* cm^{-3}).

Rates of gross and net primary production of benthic autotrophs have been measured in a handful of forests (Alongi 1989). In a detailed study under well-developed canopies of mixed *Rhizophora* forests in Australia, rates of gross primary production range from −281 to 1,413 μmol O_2 m^{-2} h^{-1} with large variations among intertidal zones and seasons (Alongi 1994). The P:R ratio in these soils range from −0.60 to 1.76 with a mean of 0.15, indicating net heterotrophy. However, on bare tidal flats adjacent to the mangroves, sufficient light is available to sustain algal mats and populations of various algal grazers (Alongi 1994). Benthic microbial mats often occur in mangroves, especially in scrub or dwarf forests, where sufficient light penetrates to the soil surface. Gross primary production ranges from 6 to 15 mmol O_2 m^{-2} h^{-1} across a range of light intensities for an average GPP of 12 mmol O_2 m^{-2} h^{-1} in algal mats among dwarf mangroves on Twin Cays, Belize (Joye and Lee 2004). These mats are also highly active sites for microbial nitrogen cycling and given their large area are likely to play a key role in nutrient cycling.

Measurements of photosynthesis by macroalgae on pneumatophores and prop roots have been made, mostly in the Caribbean; some forests accomodate quite large macroalgal communities (Burkholder and Almodóvar 1973; Rodriguez and Stoner 1990). In Colombia, prop roots of *Rhizophora mangle* are usually dominated by red algae, such as *Bostrychia calliptera, Catenella impudica*, and *Caloglossa leprieurii*. The algal associations are productive. Peña et al. (1999) measured maximum photosynthetic rates for *B. calliptera* of 126 ± 4 μmol O_2 mg chl a^{-1} h^{-1} in water and 52 ± 9 μmol O_2 mg chl a^{-1} h^{-1} in air. Slower rates of photosynthesis were measured for *C. leprieurii* of 98 ± 9 μmol O_2 mg chl a^{-1} h^{-1} and 30 ± 11 μmol O_2 mg chl a^{-1} h^{-1} in water and air, respectively. Light did not appear to be an important factor limiting primary production, but stress induced by desiccation was. In Florida forests, Dawes and his colleagues (Dawes 1996; Dawes et al. 1999) measured respiration and photosynthesis of both turf and epiphytic macroalgae on pneumatophores of *A. germinans*. Mean rates of NPP for the turf algae range from 5.8 to 10.6 mg O_2 g DW^{-1} h^{-1} in June and October, respectively, with P:R ratios ranging from 5.8 to 6.3. For the epiphytes, rates of primary production are usually higher (10–16.9 mg O_2 g DW^{-1} h^{-1}) with higher P:R ratios of 7.7 to 12.7. On an aerial basis, carbon fixation was 0.8 g C m^{-2} day^{-1} for turf algae and 2.7 g C m^{-2} day^{-1} for epiphytic macroalgae. It is unclear how these rates compare with tree production, but presumably, these communities have a significant role to play as food and refugia for consumers.

2.6 Life in the Canopy and Root Epibionts

Links between mangrove trees and animals in the canopy and on roots aboveground are extremely diverse, with close associations between tree and animal pollinators and mutual interactions of mangroves with ants, butterflies, monkeys, and birds (Ellison and Farnsworth 2001). Some interactions are highly complex, such as the indirect protection offered by ants to reduce herbivory by crabs (Offenberg et al. 2006) and direct intervention by ants to reduce the abundance of scale insects in some forests (Ozaki et al. 2000). While many such interactions and associations have been well-documented, the actual impact on canopy production or rates of secondary production of canopy dwellers is unknown.

The impact of insects within the canopy has been examined in several forests in the Caribbean and in Australia, with often variable results. In Belize, Lepidoptera larvae are the most common insect herbivores; Farnsworth and Ellison (1991) observed that these creatures can damage 4–25% of leaf area of *R. mangle* and 8–36% of the area of *A. germinans* leaves. On average, there is less damage to *Avicennia* than *Rhizophora* leaves which is a contrary result to what Robertson and Duke (1987) found earlier in northern Australian forests. In a survey of 25 mangrove species, they observed mean leaf area losses of 0.3–35% with a mean coefficient of variation of 26.6%. Only 2% of canopy production in mixed *Rhizophora* forests enters the direct herbivore pathway (Robertson and Duke 1987). Insect herbivory accounts for only a small to modest percentage loss of canopy biomass (usually <10%), but there have been some well-documented cases of defoliation of entire stands (Anderson and Lee 1995; McKillup and McKillup 1997; Duke 2002).

The differences in rates of herbivory between mangrove species and forests are most often explained in terms of differences in leaf chemistry and age. Species with thick leaves or with leaves containing chemical defenses, such as polyphenolic acids (*Rhizophora*) or poisonous latex (*Excoecaria*) presumably would be grazed less than species whose leaves have lower tannin concentrations (*Avicennia*) or higher levels of nitrogen (*Heritiera*). Differences in other factors, such as species composition and abundance of herbivore guilds, tree age, climate, and leaf mineral composition, presumably also result in differences in leaf losses (Saur et al. 1999). Why some forests are defoliated while others are not remains an open question worthy of further research.

Other invertebrates (e.g., grapsid crabs) and large mammals (deer, monkeys, cattle, hippopotami) are known to eat mangrove leaves, flower buds, flowers, and fruits directly off the tree (Barrett and Stiling 2006), but the actual amounts consumed are unknown. Herbivory by wood-borers was, until recently, unquantified. In Belize, Feller and Mathis (1997) and Feller (2002) studied the role of wood-boring insects on the canopy of *R. mangle*. Wood-borers killed over 50% of the mangrove canopy whereas leaf herbivores removed <6% of the canopy. Twig borers and stem girdlers also damage meristems and shoots that alter tree structure and form. To what extent wood-boring insects damage or kill mangrove canopies in other parts of the world is unknown, but it may be significant.

The epiphytic algae growing on prop roots and pneumatophores are a prime food source for many herbivores that are also often located on the forest floor or on the tree (Section 5.3). These migratory animals are mostly terrestrial, semi-terrestrial, and marine arthropods, and their distribution and abundance is often related to the extent of epiphytic algal cover and fine particle deposition (Proches et al. 2001; Proches and Marshall 2002). Many colonizing invertebrates are meiofaunal in size, exhibiting quite complex spatial and temporal patterns that relate not only to algal and particle cover but height of the pneumatophore and frequency of wetting.

Root epibiontic communities thrive in some mangrove forests, especially when tidal waters are sufficiently transparent to permit algal photosynthesis, colonizing space for attachment by a wide assortment of invertebrates such as sponges, bivalves, and bryozoans (Ellison and Farnsworth 2001). Most research effort into the dynamics of mangrove root epibiontic communities has focused on prop roots of *Rhizophora mangle* in the Caribbean (Bingham and Young 1995; Lacerda et al. 2002; Engel and Pawlik 2005). These communities are typically zoned in relation to tides and in responses of individual species' susceptibility to desiccation. Little work has been done on the energetics of these communities; this is unfortunate as some filter-feeders such as oysters can assimilate a significant amount of organic matter directly from the water column (see Section 5.3.4).

These variegated and highly diverse communities attain a high biomass on some prop roots and facilitate root growth and production (Ellison and Farnsworth 1992). Sponges in particular may preclude colonization of and damage to *Rhizophora* roots by isopods and may facilitate nitrogen uptake by the roots, inducing the formation of fine roots which absorb ammonium produced by the fouling communities (Ellison et al. 1996). Indeed, prop root flora and fauna are important in nutrient transformation processes. In Australian *Rhizophora* forests, the prop root epibiota take up dissolved inorganic nitrogen equivalent to 62% of nitrogen imported by tides and 60% of silicon uptake by the entire ecosystem, probably for epiphytic algal photosynthesis and bark formation (Alongi 1996).

While the bulk of mangrove canopy biomass, including above-ground root epiphytes, remains unconsumed in most forests, in Chapters 4 and 5 we will see that various organisms in tidal waters, and on and beneath the forest floor, have an important role to play in materials and energy flow. But first, it's important to understand the importance of water circulation and sediment dynamics to the energetics of mangrove ecosystems.

Chapter 3
Water and Sediment Dynamics

3.1 Introduction

The rise and fall of tides in waterways that snake through the forests is one of the most conspicuous features of mangrove ecosystems. The movements of water and sediment into and out of mangrove estuaries are facilitated by tides and, to a lesser extent, waves. Tidal and wave energy in any estuary constitutes an auxiliary energy subsidy; tides allow mangrove forests to store and pass on new fixed carbon and benefits animals adapted to make use of subsidized energy. Tides do the work of bringing nutrients, food, and sediments to mangroves and their food webs, as well as exporting waste products. This subsidy is an advantage in that organisms do not have to expend energy on these processes and can shunt more energy to grow and reproduce. For the most recent and exhaustive treatment of the role of physical processes in mangroves, I refer the reader to Mazda et al. (2007).

3.2 Tides

3.2.1 Flow in Relation to Geomorphology

Although tidal range varies greatly worldwide, the tidal circulation within most mangrove waterways is characterized by a pronounced asymmetry between the ebb and flood tides, with the ebb tide being shorter but with stronger current velocity than the flood tide. Current velocities in the tidal creek can often exceed 1 m s^{-1} but only rarely approach 0.1 m s^{-1} in the forest (Wolanski 1992). This asymmetry results in self-scouring of the tidal waterways to the extent that the bottom of most water channels are composed of rock, gravel and sand, with little or no accumulation of fine sediment.

The velocity of tidal circulation ultimately depends on the geometry of the ecosystem, especially the ratio of the forest area to the waterway area and the bottom slope of the forest (Table 3.1). In the few areas where such measurements exist, the ratio appears to be on the order of 2–10, with a very small forest slope

Table 3.1 Asymmetry of tidal flow in relation to the ratio of forest area to waterway area in some mangrove ecosystems (Data from Wolanski et al. 1980, 1990; Woodroffe 1985b; Wolanski and Ridd 1986; Wolanski 1989; Wattayakorn et al. 1990; Mazda et al. 1995; and Larcombe and Ridd 1996)

Location	Forest/ waterway area	Maximum velocity (m s^{-1}) flood	Maximum velocity (m s^{-1}) ebb
Coral Creek, Australia	5.5	1.2	1.6
Tuff Crater, New Zealand	44.0	0.4	0.6
Wenlock River, Australia	N/A	1.0	2.0
Chwaka Bay, Zanzibar	N/A	0.3	0.5
Dickson's Inlet, Australia	6.2	0.7	0.8
Klong Ngao, Thailand	2.7	0.4	0.8
Hinchinbrook Channel, Australia	2.1	0.5	0.9
Ross Creek, Australia	N/A	0.4	0.8
Fukido-Gawa, Japan	12.8	0.5	0.7

usually within the range of 1×10^{-3} to 4×10^{-3} (Wolanski 1992). The tidal prism of a mangrove estuary thus increases greatly with an increase in the ratio between forest area to waterway area.

The importance of the interaction between the geometry of tidal creeks and the mangrove forest in what causes asymmetry of tides has been determined by numerical modelling. The modelling exercise of Mazda et al. (1995) found that the dominance of the ebb tide is due to friction in the mangrove forest. The level of friction is in turn controlled by the density of the forest; inside the forest, the water level and the current velocity are strongly controlled by drag force due to the vegetation. The denser the forest, the greater the drag. This results in slower current velocity, and greater tidal asymmetry in the waterway. However, the relationship is not straightforward. Due to differences in the phases of the tide, the peak velocity in the waterway decreases at flood tide and increases at ebb tide for increasing levels of drag force, but when the drag force is excessive, the ebb flow is reduced, allowing the waterway to silt. There is therefore a natural feedback relationship among the vegetation, water, and sediments. These feedback mechanisms have great implication for human impacts on mangroves in that a reduction in the size of the forest (e.g., from land reclamation) results in a reduction of the tidal asymmetry and silting of the waterway.

This phenomenon is unique to estuaries with extensive vegetated tidal wetlands, including mangroves (Wolanski 2007). In fact, in most other estuarine waterways without extensive tidal wetlands, the flood tide is usually larger than the ebb tide (Aubrey 1986; Friedrichs et al. 1992). This is partially due to the fact that other coastal vegetation, such as salt marshes, do not provide the same level of friction on water flow. For instance, the friction coefficient used to estimate tidal flow, known as the Manning resistance coefficient, is usually within the range of 0.2–0.7 for most mangrove forests, whereas the coefficient measured in salt marshes is two to three orders of magnitude less (Friedrichs et al. 1992; but see Wolanski 2007).

An additional asymmetry of the currents in mangroves is the direction of the currents in relation to forest position. At rising tide, the currents flow into the forest perpendicular to the banks, while at falling tide they are oriented at an angle (typically 30–60°) to the bank (Wolanski et al. 1980). This lengthens the pathways of water at falling tide, reducing the chance that materials, such as mangrove propagules, can escape the forest.

Not all mangrove-fringed systems show a pronounced tidal asymmetry. In Cocoa Creek in northern Australia, Aucan and Ridd (2000) found only slight dominance of the ebb tide, as much of the tidal wetland is unvegetated mud flat which has a small friction coefficient. The primary factor reducing the ebb dominance was the very low slope of the forest surface. Water floods the vegetation at rising tide as sheet flow, but after ebb tide begins, the water that is covering the salt flats and mangroves can leave only via a very small and shallow creek which is constricted by mangrove and marsh vegetation. This leads to increased friction and a delay in water leaving the forest and salt flats.

The drag forces and subsequent time delay in the movement of tidal water within the forests and adjacent waters results in another characteristic feature of mangroves: the mixing and lateral trapping of water. Trapping of tidal water has been observed in a number of mangrove estuaries, and was first observed in salt marsh estuaries, where the effect is lessened by short marsh grass (Wolanski 2007).

Lateral trapping of water within the forest is a dominant process controlling longitudinal mixing in mangrove waterways (Wolanski and Ridd 1986). The trapping phenomenon occurs when some of the water flowing in and out of an estuary is temporarily retained in the mangrove forest to be returned to the main water channel later. Trapping of water is enhanced in the dry season when there is little, if any, freshwater to cause buoyancy effects on water circulation. In the wet season, the buoyancy effect is important as freshwater is trapped in the forest at high tide, and as a floating lens or boundary layer hugging the river banks at low tide. This effect means that the forests control the runoff of freshwater, especially at the end of a flood. In the Klong Ngao mangroves of Thailand, waters are similarly well-mixed as are those in the Australian estuaries with a positive salinity gradient (increase in salinity towards the mouth). The situation reverses in the long, dry season where salinity increases landward; this salinity change is probably driven by evapotranspiration from the forests. So, in essence, the evaporation of water and buildup of salt generated by the physiological activities of the trees helps to generate gradients of salt and other materials, both laterally and longitudinally. The trapping of mangrove and coastal waters has also been observed in Gazi Bay in Africa where additional physical forces of shoreward winds generate shoreward currents over a coral reef flat (Kitheka 1996). Thus, tides, onshore winds, and waves can operate synergistically to trap river waters in the mangrove waterways and along the coast.

A significant lateral gradient within mangrove creeks during the dry season can be attributed to high evapotranspiration. Weak stratification usually prevails at the headwaters of mangrove-fringed tidal waterways in the wet season as the result of freshwater input from rivers. Similar gradients in salinity and slow flushing times have been found in the Konkoure River delta in Guinea (Wolanski and Cassagne 2000). In arid-zone mangrove estuaries, the salinity structure is inverse, due to the lack of freshwater input and the high evaporation rate, especially in relation to the salt flats and mangroves bordering the estuary (Wolanski 1986; Ridd and Stieglitz 2002); salinities can be extreme, often >50.

The behavior of tidal water is also longitudinally complex. Longitudinal diffusion is proportional to the square of the water velocity, which means that at the headwaters of mangrove creeks where currents are very small, mixing rates are also very small (Ridd et al. 1990). Along the length of a waterway, water speed decreases from the mouth to the headwaters. The longitudinal (and cross-sectional) gradients in current speed are partly the result of shear dispersion processes which are magnified by the presence of the forest; this diffusion process drives the intensity of mixing and trapping. All of these complex processes translate into residence times for water near the head of a mangrove waterway that are long, especially in the dry season. This has direct biological and chemical consequences as contaminants introduced at headwaters may be retained longer than if they were introduced further downstream (Wolanski et al. 2000).

To add to the physical complexity of water circulation, all estuaries, including those inhabited by mangroves, exhibit secondary circulation patterns superimposed on the primary tidal circulation (Nunes and Simpson 1985). This phenomenon is responsible for the often observed trapping of floating mangrove debris, including propagules, in density-driven convergence fronts during a rising tide (Stieglitz and Ridd 2001). This secondary circulation has been measured in detail in the Normanby River in northeastern Australia (Ridd et al. 1998). A well-developed axial convergence was found in the estuary, unbroken for 30 of the estuary's 80 km length. These fronts occur in well-mixed estuaries due to the interaction between the velocity of water across the estuary and the density gradient up the estuary; due to friction, the velocity is slower near the river banks than in the center of the estuary thus causing, on flood tides, a greater density mid-channel than at the banks. The water in the center of the estuary sinks causing a two-cell circulation pattern. There is a swift breakdown of the convergence when the ebb tide begins. There are clear biological implications in the existence of these cells, and based on their experiments, Ridd et al. (1998) and Stieglitz and Ridd (2001) made the following observations:

- A net upstream movement of floating debris (e.g., propagules) occurs, on the order of several kilometers per day.
- When the secondary cells are present, propagules are unlikely to enter the mangrove forests within the estuary.
- The propagules accumulate in large numbers in 'traps' upstream from the convergence and upstream from the mangrove fringe.
- Trapping of the propagules upstream is not conducive to the natural strategy of seed dispersal of mangroves.

3.2.2 Flow in Relation to Vegetation and Other Biological Structures

The presence of trees, roots, animal burrows and mounds, timber, and other decaying vegetation lying on the forest floor exert a drag on the movement of tidal waters within a forest. The drag force of the trees can be simplified to a balance between the slope of the water surface and the flow resistance due to the vegetation. Water flow in the forest depends of the volume of the trees relative to the total forest area, or the effective vegetation length scale, L_E (Mazda et al. 1997b), $L_E = (V - V_M)/A$, where V = volume of an area covering at least one tree (plus its roots above-ground) over the substrate and a height above the substrate. This can be imagined as a box sitting on the forest floor with a tree growing out of it, with dimensions of length × width X height (or depth). In the equation above, A = the total projected area of the vegetation in the box (V), and V_M = the total volume of the vegetation in the box (V). L_E values are very much dependent on the height (or depth) of the box. For example, at a high tide in a forest of >0.5 m in water depth, L_E approaches 1.0 m, nearly equal to the average spacing between trees; when depth is <0.1 m, L_E ranges from 0.15–0.25 depending on mangrove species.

The flow resistance can be approximated by a drag coefficient, C_D, which can vary in mangrove forests from 0.4 to 10 (Mazda et al. 1997b, 2005). The drag coefficient varies with the Reynolds number, $R_E = uL_E/\nu$, where L_E is defined above, u = depth-averaged tidal water velocity (unidirectional) and ν = kinematic viscosity, which is the ratio of the dynamic viscosity (a measure of the tendency of the seawater to stick to itself) to the seawater density. Mazda et al. (1997b, 2005) measured and calculated both the drag coefficient and the Reynolds number for a variety of mangrove species. They found a unique relationship (Fig. 3.1) in which the value of the

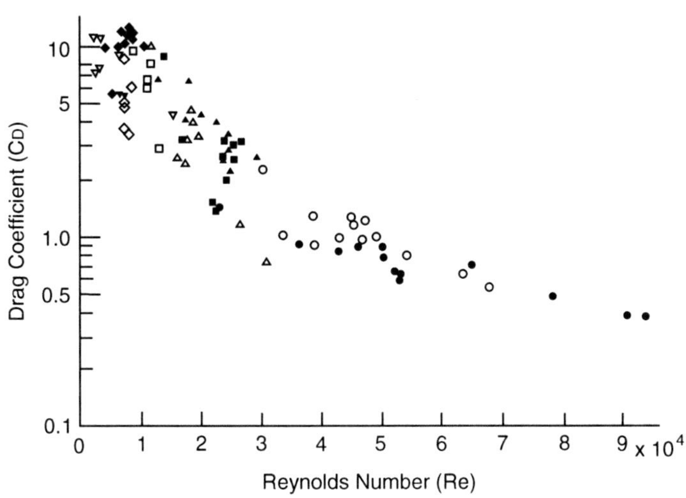

Fig. 3.1 Relationship between the drag coefficient and Reynolds number as calculated for mangrove forests in Australia and Japan (Modified from Mazda et al. 1997b)

drag coefficient decreases with increasing values of the Reynolds number. From a biological perspective, it is important to note that for Reynolds numbers $>5 \times 10^4$ the drag coefficient converges towards a value of 0.4 which is the magnitude of flows around a single cylinder (a single tree trunk!). Conversely, at low Reynolds numbers, the drag coefficient approaches 10. This means that in densely vegetated forests in shallow water (typically <1 m depth) during tidal inundation, prop roots and pneumatophores play an important role in the flow of tidal water in the forest. It is important to note that the calculated Reynolds numbers are generally large ($X\ 10^4$) meaning that in mangrove forests inertial forces (momentum) is relatively more important than viscosity. That is, the momentum of tidal forces is greater than the shear stress induced by the presence of obstacles, including friction with the soil surface. These results show that the dynamics of tidal waters in mangrove forests change in relation to tree species, density of the vegetation, and state of the tides.

The effect of the vegetation is somewhat more complex, in that currents in the forest itself are not negligible and a secondary circulation pattern is usually present due to the density of the vegetation and the overflow of water into the forest at high tide. This secondary circulation enhances the trapping effect of tides (Mazda et al. 1999). Using numerical models, Mazda et al. (1999) found that the drag force has two main influences: (1) inundation of the forest is inhibited, and this decrease in water volume results in smaller dispersion, and (2) the trapping of water in the forest is enhanced, favoring dispersion. These effects are, however, non-linear, with minimal dispersion at intermediate drag force, but large dispersion with no and high vegetation density. Simply, the magnitude of tidal trapping depends on the drag force due to the vegetation, so the magnitude of dispersion depends ultimately on the vegetation density. The resistance offered by mangrove trees to water flow has been experimentally tested in a flume (Struve et al. 2003). Drag coefficient values commonly range from 0–4.5, with the drag force increasing with increasing tree surface area and tree density, as predicted by the models used by Mazda and his colleagues.

Animal structures also impact on water circulation in mangroves (Ridd 1996; Stieglitz et al. 2000a, b; Heron and Ridd 2001, 2003, 2008; Susilo and Ridd 2005; Susilo et al. 2005). Crabs and other benthic organisms produce numerous burrows and other biogenic structures in the mangrove forest floor through which tidal waters flow. In initial experiments, Ridd (1996) found that tidal waters flow through a labyrinth of interconnected burrows in the same direction as the surface current. The flow through the tubes is caused by a pressure difference between multiple burrow openings, with flow velocities of up to 30 mm s^{-1}. Using a conservative estimate for burrow density, the total quantity of water that flows through burrows in a 1 km^2 area of forest can range from 1,000 to 10,000 m^3, representing from 0.3% to 3% of the water volume moving through a forest. In further experiments using salt as a tracer, Ridd and his colleagues measured passive irrigation through crustacean burrows (Steiglitz et al. 2000a, b). Of biological importance was that parts of the burrow were no further than 20 cm apart, reducing the diffusion distances below-ground of salt excluded from mangrove roots. Thus, well-flushed burrows are an efficient mechanism by which salt can be transported away from roots. For most tree roots, diffusion of salt to a burrow will be more efficient and faster than diffusion to the soil surface.

More sophisticated fluid dynamics modelling indicates that the flow of water through animal burrows is greatly influenced by burrow architecture and depth, slope of the forest floor, the number of loops in the burrow, and location of roots relative to the burrow (Heron and Ridd 2001, 2003, 2008). Flushing times for a 1.2 m-deep burrow with the location of a mangrove root between burrow openings and downstream of the burrow were 5 and 28 min, respectively. The flushing times for these two root locations lengthened to 15 and 38 min, respectively, when a second burrow was added into the model computations. All flushing times estimated from the model were consistent with the idea that burrows are well-flushed over a single tide. Multiple-loop burrows are also flushed over a single tide, but modeling shows that flushing is enhanced as upper loops can be completely drained, increasing the volume of water flushed through the structure. However, there may be a depth limit to flushing for some loops, on the order of 30–35 cm, depending on loop complexity and angle relative to the soil surface (Heron and Ridd 2008).

The transport of salt derived from the tree roots results in variations in the density of water flowing through the burrows, having an impact on flushing time. Heron and Ridd (2003) found that with higher water density, flow is limited or even halted. As the model assumes that the burrow is empty, in reality, the presence of an organism would lengthen the actual flushing time (assuming that the animal is not actively irrigating its burrow). Burrows, tubes, mounds, and other biogenic structures impart some significant delay in water flow, assisting in the trapping of water in the forest. This impact also has implications for solutes carried by tidal waters, especially when these structures are inhabited. For instance, the mudskipper *Periophthalmodon schlosseri* stores air in its burrow (Ishimatsu et al. 1998). This is a behavioral pattern as mudskippers commonly inflate their buccopharyneal cavity before entering a burrow and exhale inside the burrow. The accumulation of air inside the burrow provides a significant oxygen supply for eggs and developing embryos as mudskippers commonly deposit their eggs on the ceiling of the burrow, forming a spawning chamber. Such behavior has biogeochemical implications, influencing rates and pathways of microbial metabolism in the soil (Section 5.4).

3.3 Groundwater

Water derived from land often takes a subterranean path, so given their position between land and sea it is not surprising that mangrove forests can have significant groundwater flow. This flow can be an adaptive advantage in that groundwater is likely to be an important pathway for the removal of salt excluded by tree roots (Ridd and Sam 1996; Sam and Ridd 1998) and reduced by-products (e.g., sulfides, methane) of microbial decomposition of organic matter (Ovalle et al. 1990). Given the often high silt and clay content of mangrove soils, it is very likely that crab burrows, fissures, and cracks in the substrate facilitate the movement of groundwater (Wolanski 1992).

The flow of groundwater in mangrove systems usually has three components (Mazda et al. 1990; Mazda and Ikeda 2006):

- A near-steady flow towards the open sea due to the pressure gradient induced by the difference in height between water levels in the forest and the open sea.
- A reversing tidal flow with a damped amplitude and delayed phase towards the forest.
- A residual flow towards the forest caused by the damped tidal flow. This residual flow reduces the outflow of water from the forest towards the sea.

The flow of groundwater due to differences in the water table between the groundwater in the forest and the creek, and via tidal flushing of animal burrows, may be equivalent (Susilo et al. 2005; Mazda and Ikeda 2006).

Evidence of the impact of groundwater flow often comes from the occurrence of vertical salinity anomalies (salinity differences between ebb and flood tides) and the presence of saline groundwater in wells adjacent to mangroves (Kitheka 1998). In mangrove-fringed Mida Creek in Kenya, Kitheka (1998) found a low salinity lens of water at the bottom of the water-column, beneath water of higher salinity. During droughts, groundwater seepage slows and hypersaline conditions occur, especially in backwater regions with restricted circulation and high rates of evaporation. In Mida Creek, however, flood tidal speeds are greater than ebb currents, so groundwater flow to the sea is limited.

The influence on groundwater-derived nutrients on nutrient dynamics in mangrove waterways can be significant, and can often vary depending on the season (Kitheka et al. 1999). This is best seen in the seasonal differences in the groundwater and tidal flux of ammonium, nitrite + nitrate, and silicate (Table 3.2) in Mida Creek. Groundwater flow is greater in the wet season for silicate, whereas the opposite is true for $NO_2^- + NO_3^-$, with little difference for NH_4^+. In both seasons, flood tide is usually greater than ebb tidal flow. Nutrient concentrations are greater in the groundwater than in the creek, and with the exception of $NO_2^- + NO_3^-$ in the groundwater, are greater in the wet than in the dry season.

Table 3.2 The flux (and concentration) of nutrients from ebb and flood tides plus the contribution of groundwater input (GW) in the dry and wet seasons at Mida Creek, Kenya (Kitheka et al. 1999). Flows are in g $s^{-1}$and concentrations in parenthesis are μM. Values for nutrient concentrations are the average of both tides

Nutrient and water state	Dry season	Wet season
NH_4^+GW	0.2 (8.0)	0.3 (11.05)
Ebb	8.7 (0.22)	15.1 (0.39)
Flood	11.7	20.2
$NO_2^- + NO_3$ GW	28.3 (1,124.1)	16.2 (642.87)
Ebb	59.7 (1.52)	103.3 (2.64)
Flood	80.0	138.4
Si $(OH)_4^+$GW	7.6 (149.8)	23.8 (470.25)
Ebb	499.6 (6.35)	1,254.2 (15.94)
Flood	669.1	1,680.0

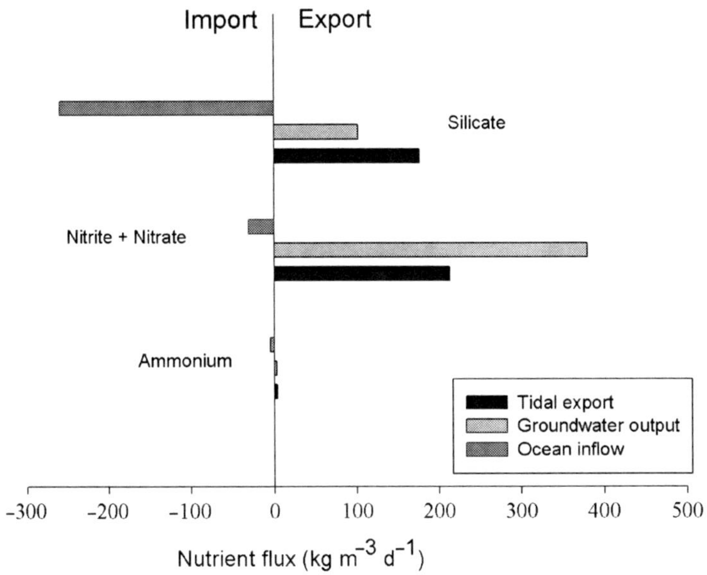

Fig. 3.2 The relative importance of groundwater versus tidal export and ocean inflow in Mida Creek, Kenya (Data from Kitheka et al. 1999)

The data in both Table 3.2 and Fig. 3.2 show that groundwater seepage contributes between 8–140% of the net ammonium and nitrite + nitrate flux, but <5% of the net silicate flux. At a systems-level, tidal export is less than groundwater and oceanic inputs, showing that there are more nutrients imported from the sea and land than exported by tides. Mida Creek is therefore a source of nitrite + nitrate, but a sink for silicate and ammonium. Most of the $NO_2^- + NO_3^-$ enters the creek through groundwater and only a very small amount is imported from the open sea. For silicate, a large percentage enters the creek from the open sea although the groundwater supply is still large. This creek is somewhat anomalous compared to most other creeks as it has no significant river drainage.

In most other mangrove creeks, nutrient flows are often dominated in the wet season by inputs from surface water runoff and rivers (Kitheka 1998; Drexler and DeCarlo 2002). A variety of factors come into play in terms of the relative contribution of groundwater to total water flow. On islands in the Federated States of Micronesia, Drexler and DeCarlo (2002) traced the various sources of freshwater and seawater in both river-dominated and interior mangrove ecosystems using chloride as a chemical tracer. In the river-dominated mangroves, the average groundwater contribution was 5% with a much greater contribution of 20% for the interior site closer to land, with a hydrologic connection between mangrove forest and an upstream freshwater swamp. Greater groundwater input at the interior site demonstrates that such connections help to alleviate salinity stress and possible desiccation of the forest. Groundwater seepages, although usually small to modest, may in fact promote the growth of mangroves and associated flora in areas where growth may be otherwise unsuitable, by reducing anoxia and providing a crucial input of freshwater to dilute salt.

3.4 Waves

Although mangroves grow best in quiescent conditions, they often receive a substantial amount of wave action, especially when growing on the fringes of open bays and estuaries. Mangrove forests attenuate wave energy, but only recently have laboratory and field experiments been conducted to understand the factors responsible for the dissipation of waves (Brinkman et al. 1997; Mazda et al. 1997b, 2006; Massel et al. 1999; Quartel et al. 2007; Vo-Luong and Massel 2008). There are two primary mechanisms responsible for wave dissipation in mangrove forests: (1) multiple interactions of waves with mangrove trucks and roots and (2) bottom friction.

The coefficient necessary to calculate bottom friction accurately is not known, so attention has focused on the effect of the presence of trees and their roots. Forces induced by waves on tree stems and roots are inertial and drag-type forces, with drag force dominating for most mangroves. With increasing tree diameter and density, the degree of wave attenuation increases. However, interactions between tree stems can be expected to influence the extent of drag. Using a discrete vortex method, Massel et al. (1999) modified the drag coefficient to be dependent on the density of tree stems. Waves within a mangrove forest are strongly dissipated by these interactions. In Tong King delta, Mazda et al. (1997a) found that seedlings of *Kandelia candel* were not effective in dissipating wave action, but larger trees of the same species reduced wave action by 20% over 100 m distance. Therefore, dissipation of wave energy is a function of total tree area which is in turn a function of both tree diameter and forest density.

Water depth and the spectral characteristics of waves also play an important role in energy dissipation (Vo-Luong and Massel 2008). Figure 3.3 illustrates how the density of the forest and the distance from the forest front facing the sea results in the reduction in waves of 2–3 s period (i.e., small, choppy waves). For a very dense forest, wave energy is almost totally dissipated within the forest, within 40–50 m from the mangrove/sea boundary. In a low density forest, about 35% of the incident wave energy is still extant behind the forest area. Empirical data for the experiments carried out by Brinkman et al. (1997) are in overall agreement with the modelling results, except that attenuation in real forests is not as abrupt; about 20% of the incident wave energy still remains after traversing 250 m into the forest. However, larger, swell waves (8 s period) are not dissipated as rapidly as small waves (Fig. 3.4). Larger waves propagate several hundreds of meters into mangrove forests, with the distance depending partly on tree species. Thus, the extent of wave propagation depends very much on the wave period.

Tsunamis, however, are very different from wind waves and tidal waves (Latief and Hadi 2007), having periods usually between 10 min and 1 h, and propagating like a tidal bore in that its momentum increases as it moves towards a shoaling coast. Models developed in Japan to predict the attenuation of tsunami energy by mangrove forests suggest a 90% reduction in maximum tsunami flow pressure for a 100 m wide forest with a tree density of 3,000 stems ha^{-1} (Hiraishi and Harada 2003). Model results obtained for various types of vegetation, including mangroves, were similar (Hamzah et al. 1999; Harada and Imamura 2005; Latief

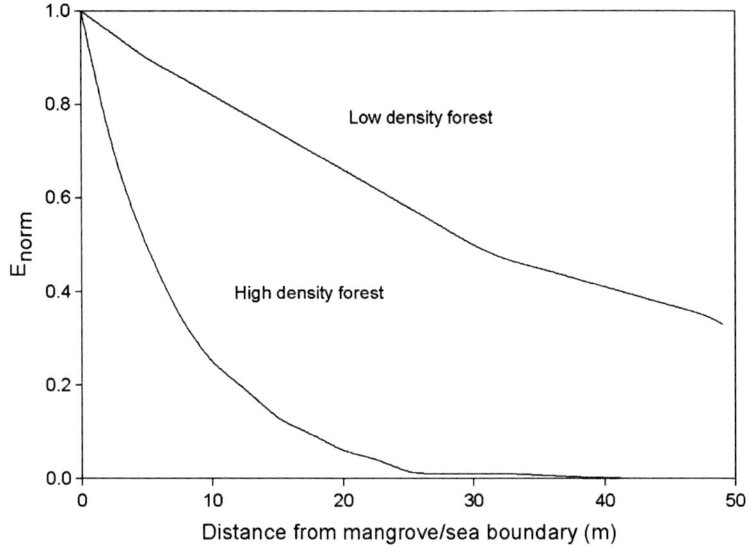

Fig. 3.3 Influence of stand density on the attenuation of wave energy in tropical mangrove forests, as modeled by Massel et al. (1999). E_{NORM} is the calculated normalized wave energy which is the ratio of wave energy at a distance x from the mangrove/sea boundary and the incident wave energy

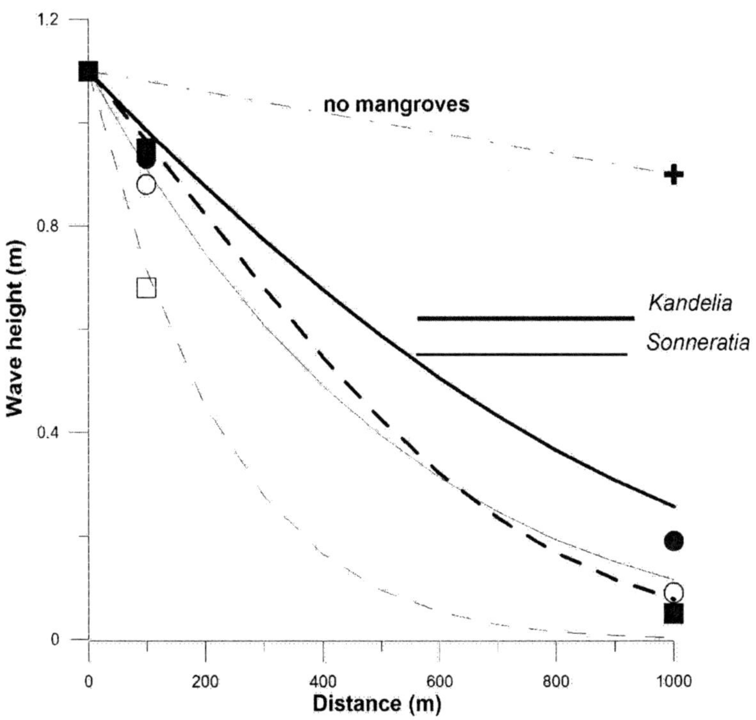

Fig. 3.4 A model showing how swell waves (8 s period) propagate several hundreds of meters into *Kandelia* and *Sonneratia* forests (Adapted from Barbier et al. 2008)

and Hadi 2007; Tanaka et al. 2007), but also suggested species differences in the ability to attenuate the energy of a tsunami. The extent to which mangroves provide protection from tsunamis depends on: width of the forest, tree diameter and species, tree density, soil texture, forest location, presence of foreshore habitats, size and speed of the tsunami, distance from the tectonic event, angle of tsunami incursion relative to the shoreline, slope of the forest floor, and the proportion of above-ground biomass vested in roots (Alongi 2008).

Mangrove forests can offer limited protection to the coastline (and human settlements). However, even emergent mangroves need protection. As saplings are not effective wave attenuators, protection is needed for the development of mangrove plantations. The use of bamboo poles in front of a young mangrove plantation can be an effective and inexpensive way to shield mangroves from wave disturbance (Halide et al. 2004). Bamboo poles of 8 cm diameter placed in front of mangroves at a density of 1–4 poles m^{-2} can attenuate 50% of incident wave energy. Such protection is often necessary for initial stages of mangrove reforestation projects.

3.5 Sediment Transport and Flocculation

As the bamboo experiment amply demonstrates, mangroves grow best in quiescent conditions where wave energy is low. Tidal flow and the subsequent attenuation of water flow within the forest results in the deposition of fine particles from the overlying water column. The transport of suspended sediment in mangroves is controlled by several interrelated processes (Wolanski 1995):

- Tidal pumping
- Baroclinic circulation
- Trapping of small particles in the turbidity maximum zone
- Flocculation
- The mangrove tidal prism
- Physiochemical reactions that destroy flocs of cohesive sediment
- Microbial production of mucus

The relative importance of these processes is site-specific. For example, where mangroves constitute only a narrow fringe along the shoreline, it is unlikely that they play an important role in sediment dynamics (Bryce et al. 2003). At the other extreme, where mangrove forests are extensive in relation to waterway area, we can expect mangroves to have a profound impact of the transport and deposition of fine particles.

Other factors play a role in sediment transport, such as the extent of mangrove destruction, while others such as freshwater floods, appear to play a minor role. In tidal waterways draining degraded mangrove forests, the magnitude but not the direction of sediment transport is affected (Kitheka et al. 2003). In a degraded Kenyan ecosystem, there was still a net import of sediment; however, with distance from the sea, sediment trapping efficiency declined from 65% to 27% (Kitheka et al. 2003).

Sediment accretion rates were high, ranging from 0.25–3.5 cm year^{-1} owing to enhanced erosion of sediment in the degraded forest.

Most sediment imported into the forest occurs during the wet season, which is the period when riverine sediment inflow is at its height. In Kenya, net import occurs mostly during periods of high river discharge on both neap and spring tides; in the dry season, import is smaller as it occurs only on a spring tide (Kitheka et al. 2003). Similar patterns have been observed in some northern Australian mangrove estuaries (Larcombe and Ridd 1996; Bryce et al. 1998).

Secondary circulation, such as that driven by stratification in salinity (and density) or suspended sediment concentration, can sort particles by size. In the South Alligator River in northern Australia, for example, it is common along meanders to find that clay and silt particles have aggregated on the river bank leaving behind only gravel and sand on the river bed (Wolanski et al. 1988). The tidal prism is significantly increased at spring tide by the presence of extensive mangrove forests. This phenomenon in turn modifies the hydrodynamics of the tide from a flood-dominant to an ebb-dominant regime at spring tide. The system then becomes self-scouring, maintaining a deep channel in the waterway (Wolanski 1992; Bryce et al. 2003).

Along the river's edge, mud banks form not just as the result of baroclinic circulation, but also as the result of tidal pumping and mixing, especially in the turbidity maximum zone (Fig. 3.5). A turbidity maximum zone is formed within an estuary where the residual inward bottom flow meets the outward river flow. This

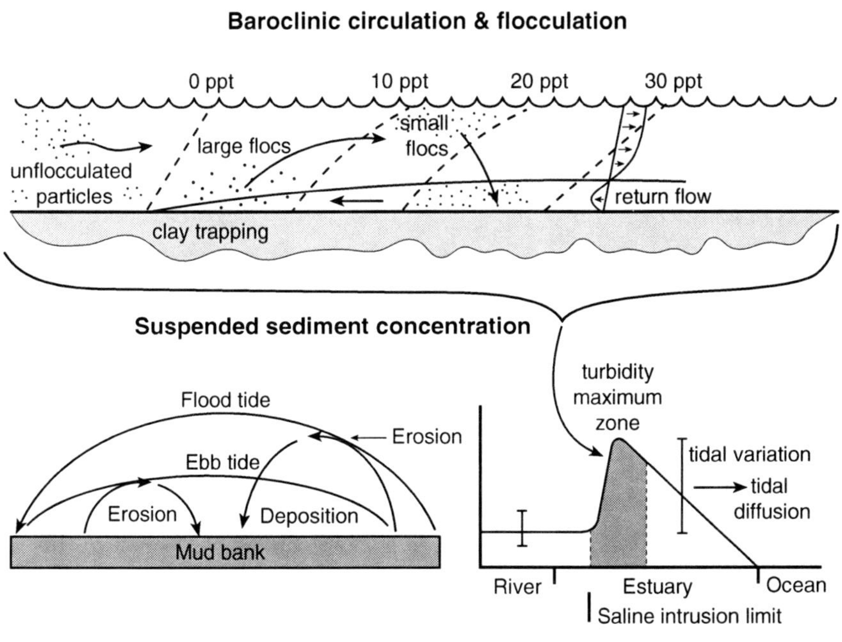

Fig. 3.5 Diagram of the effect of baroclinic circulation, tidal pumping, mixing, and flocculation in the turbidity maximum zone of a model mangrove estuary (Modified from Wolanski 1995, 2007)

zone is usually at the most landward point reached by the saline water flow. The water is generally shallowest at this point because this is a convergence point where sediment accumulates. The residual flow is seaward upstream of this convergence point and seaward in surface waters downstream of the convergence point. Thus, this is the convergence point for suspended sediment, the bulk of which saltates along the bottom (Eisma 1998). Some suspended matter settles here as the net current velocities in the zone are low. The zone is not stationary, but moves with the ebb and flow of the tide.

In relation to the turbidity maximum, flocculation of particles begins at salinities often <1; the largest flocs remain near the river bottom. The small flocs and unflocculated particles move further downstream (Fig. 3.5) with the currents, where they aggregate with local particles. As the floc size increases, they move toward the river bed when they are entrained upstream by the baroclinic circulation. Due to tidal pumping, these flocs are carried further upstream at flood tide than downstream at ebb tide. In some mangrove estuaries, such as in the Fly River delta in Papua New Guinea, there is selective trapping of clay instead of silt particles in the turbidity maximum zone (Wolanski and Gibbs 1995). Clay particles dominate the sediment fraction of the sediment along the estuary, selectively entering the mangroves and accounting for as much as 50% of the suspended particles at flood tide in the turbidity maximum zone. These particles, however, are often re-suspended by tides and wave-generated turbulence (Wolanski 1992).

The flocs are a loose matrix of clay and silt particles, typically a few tens of micrometer in diameter, with their size controlled somewhat by the strength of the tidal currents. Disaggregation starts when tidal velocities exceed $1\,m\,s^{-1}$ (Wolanski and Gibbs 1995). During spring tides, flocs are typically between 15–40 μm in diameter, and are larger (frequently >100 μm) during neap tides. The flocs are also composed of bacteria, protists, and fungi, and their extracellular products, such as mucus and threads. These microbial hitchhikers help to cement the flocs and to maintain size when subjected to turbulence.

Within the forest, turbulence is also generated by flow around the trees, resulting in flocs remaining in suspension. These flocs are cohesive, being composed of clay and silt, and form larger flocs. The settling of sediment particles in the forests occurs for a short period (<30 min) when the tides turn from rising to falling and the waters become quiescent. Settling is facilitated also by the sticking of microbial mucus and by pelletization by invertebrate excreta. Wolanski (1995) described these processes for particles in Coral Creek in northern Australia, where he observed that large quantities of non-flocculated particles entering the mangroves at flood tide are re-exported on the ebb tide, but stick to mucus at the water surface. Mucus is common in mangroves, being found on rotting tree trunks and leaves, on the sediment surface and in the density-driven lines of organic material, as described by Steiglitz and Ridd (2001).

It is therefore correct to state that by a variety of mechanisms, mangroves are not just passive importers of fine particles, but actively capture silt, clay, and organic matter (Furukawa and Wolanski 1996; Furukawa et al. 1997). The size, shape, and distribution patterns of trees have a profound impact on sedimentation. Large trees with complex root systems (e.g., *Rhizophora*) facilitate the deposition of particles to a much greater degree than trees that are smaller and of much simpler architecture (e.g., *Ceriops*). Deposition of particles occurs around slack high water when currents are minimal. The flocculation of particles results in faster settling velocities;

Furukawa and Wolanski (1996) found that most flocs settled within 30 min just before slack high tide. Until slack water, turbulent wakes created by tree trunks, roots, and pneumatophores maintain particles in suspension. However, once in the forest, conditions are unfavorable for them to be re-suspended, as the high vegetation density inhibits water motion. In Coral Creek, Furukawa et al. (1997) estimate that 80% of the particles brought in at spring flood tides are retained within the forest, corresponding to a sedimentation rate of 10–12 kg sediment per meter of creek length per spring tide. This is equivalent to a rise of sediment of about 0.1 cm $year^{-1}$ within the forest.

3.6 Sedimentation and Accretion: Short-Term Versus Long-Term Dynamics

As noted by Alfred Russel Wallace (see introductory remarks, page 1), one's first impression of mangroves is that the river banks inhabited by the forests are accreting. This observation has been made many times in the literature, but relatively few empirical measurements have been made of sedimentation processes in mangrove forests. Sedimentation rates, measured by radiotracers and by short-term measurements of changes relative to sea-level datum, do indeed show a pattern of net sedimentation. Owing mostly to different methods and regional differences in climate, the rates vary widely, from <1 mm $year^{-1}$ to more than a few centimeters per year (Table 3.3). One generalization that can be made is that sedimentation rates appear to be highest in mangroves lining rivers with high rates of freshwater discharge and in mangroves of highly impacted rivers such as those in southern China (Alongi et al. 2005b). Lowest accumulation rates occur most often in fringing mangroves bordering open bays and estuaries in the dry tropics. A variety of measurements using radiotracers in the rivers of southern Papua New Guinea, confirm the former supposition (Brunskill et al. 2004; Walsh and Nittrouer 2004). Measurements made in south Florida and in the wider Caribbean confirm the latter (Lynch et al. 1989).

These empirical measurements do not reflect the long-term dynamics of sedimentation in mangrove forests. Over decades and centuries, coastal change is very dynamic, especially in the wet tropics; sediments deposited rapidly in one mangrove forest are likely to be sediments eroded and transported from another mangrove stand upstream. An extreme example is the sedimentary environment of the coasts of the Guineas in South America (Fromard et al. 2004). Using historical photographs and remote sensing techniques, Fromard et al. (2004) showed that the mangrove dynamics over the 1951–1999 period consisted of altering periods of net accretion (1951–1966) and erosion (1966–1991), followed by a still developing, accretion phase.

These patterns, coupled with an understanding of forest processes (Fromard et al. 1998; Proisy et al. 2000, 2002), have allowed the development of a global model of mangrove and shoreline evolution (Fig. 3.6). The model shows how the development of mangroves and shorelines are closely linked. The development of a mud bank is the first phase of the cycle, as colonization by mangrove propagules takes place when the bank is stabilized above mean sea level, and subject to tidal cycles of

Table 3.3 Rates of sedimentation (mm year^{-1}) in various mangrove forests (Data from Spackman et al. 1964, 1966; Bird 1971; Woodroffe 1981; Lynch et al. 1989; Ellison 1993, 2005; Parkinson et al. 1994; Allison et al. 1995; Cahoon and Lynch 1997; Furukawa et al. 1997; Wolanski et al. 1998; Saad et al. 1999; Smoak and Patchineelam 1999; Allison and Kepple 2001; Alongi et al. 2001, 2004a, 2005b; Anthony 2004; Brunskill et al. 2004; Walsh and Nittrouer 2004; and McKee et al. 2007)

Location	Sedimentation rate
Sepetiba Bay, Brazil	1.2–1.3
Coral Creek, Australia	10
Fly Delta, Papua New Guinea	15–44
Twin Cays, Belize	0.7–1.6
Bay of Bengal	≤5
Amazon delta front	2.4–20
Purari Delta, Papua New Guinea	13–72
Ajkwa estuary, Papua	0.6–5.5
Victoria, Australia	2.3–8
Sawi Bay, Thailand	10–12
Matang Forest Reserve, Malaysia	10–31
Kuala Kemaman Forest Reserve, Malaysia	10.6
South Florida	0.4–1.1
Grand Cayman Island	1.0
Bermuda	1.0
Sherbro Bay, West Africa	1.1–1.3
Jiulongjiang estuary, China	13–60

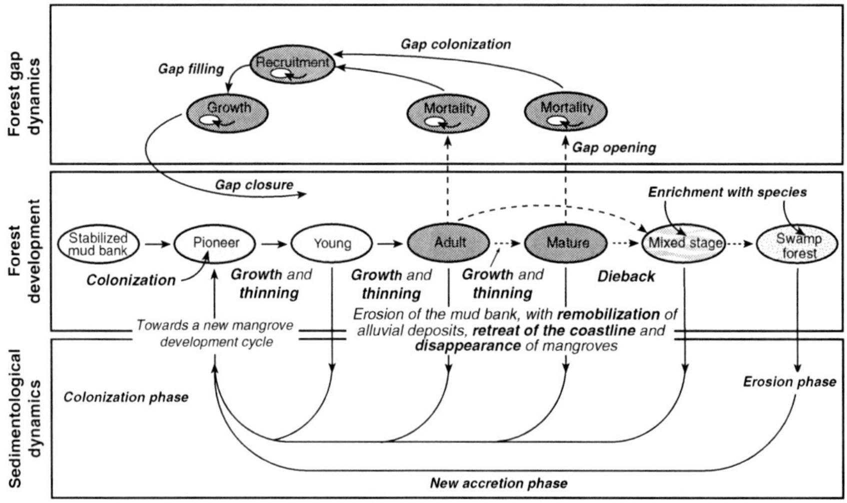

Fig. 3.6 A global model of mangrove and shoreline evolution (Modified from Fromard et al. 2004)

immersion and emersion. Over time, these pioneers grow into a young forest; stand development takes place in which competition for space and light leads to changes in species composition, size structure and density (the "self-thinning rule"), locally modified when mortality gives way to colonization of gaps in the canopy. An erosion phase may occur at any stage in this development, for example, due to storms or variations in river discharge. This is followed by an accretion phase—sediments that are remobilized by erosion are transported and eventually re-deposited in an area where tides and currents (whose patterns are altered by the regression of the shoreline) permit settlement and stabilization of a new mud bank, which is colonized by mangroves. Naturally, this is an idealized model developed from the highly dynamic area of the Amazon. Nevertheless, the model shows how mangrove forest development is married to the long-term dynamics of shoreline development. Other large-scale and long-term studies suggest that this model is generally appropriate to explain the dynamics of mangrove development on the coast of West Africa (Bertrand 1999; Anthony 2004) and southern Papua New Guinea (Walsh and Nittrouer 2004). The idea that mangrove forest structure reflects historical responses to changes in shoreline geomorphology is compatible with this model (Alongi 2008).

On a more localized scale, deposition of sediments and associated organic matter must ultimately be considered in relation to changes in sea-level, as shoreward development of mangroves is limited by the extent of tidal penetration (Woodroffe 1992). That is, for a given area of mangrove forest, net deposition cannot continue indefinitely. For example, an empirically measured sedimentation rate of 1 cm $year^{-1}$—regardless of methodology—is only a "snapshot" in time. Over long stretches of time, as sediments accrete, the actual rate of deposition will decline as the bank is inundated less frequently by tidal waters carrying fine particles. Indeed, in both Chinese and Papua New Guinean forests, rates of sediment accumulation decline with distance from the shore (Walsh and Nittrouer 2004; Alongi et al. 2005b); deviations from this pattern can be attributed to localized features, such as erosion generated by ship traffic and local instability of mud banks. Sediment accumulation rates in these regions appear to exceed local sea-level rise, but in many tropical river deltas, islands or river banks that are rapidly accumulating sediment are usually situated close to islands or banks that are eroding, suggesting a high rate of local sediment transport, but probably little net flux overall.

Over the scale of an entire delta, net accumulation is therefore likely to be smaller than estimated by extrapolation from a few samples. For example, sedimentation rates in mangroves of the Matang Mangrove Forest Reserve in peninsular Malaysia suggest that over the past century, intertidal area has expanded by about 2,500 ha (Alongi et al. 2004a). However, based on detailed surveys since the inception of the reserve, the ecosystem has actually gained only ≈1,500 ha of forest, of which 228 ha has accreted in the past decade. It is therefore reasonable to conclude that intertidal banks undergo periods of erosion and deposition, with some sediment within the system being transported locally, rather than being "new" sediment imported from sea or land. Nevertheless, the overall patterns of accumulation indicate that mangrove-lined river deltas in most tropical regions are currently growing seawards (Woodroffe 2003).

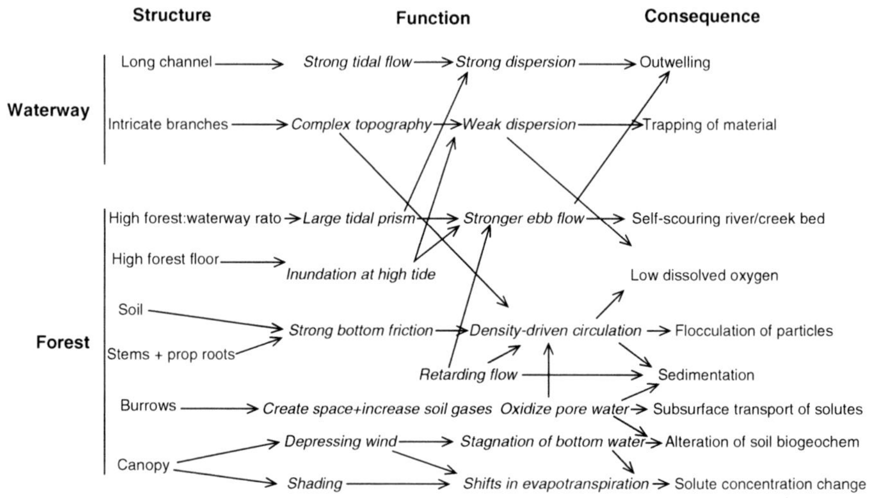

Fig. 3.7 The consequences of physical structure and function on the chemistry and biology of mangrove ecosystems (Modified from Wolanski 1992)

3.7 Chemical and Biological Consequences of Water and Sediment Flow

The flow of water and sediment in mangroves naturally has important consequences for the chemistry and biology of the ecosystem. A modification of Wolanski's (1992) diagram provides a succinct picture of how water and sediment movements impact on the biogeochemistry and ecology of the forest and waterways (Fig. 3.7).

Both the topography and morphology of the forest and its waterways, and the tidal regime, control the extent of tidal mixing and trapping, including facilitating tidal asymmetry. These factors, in turn, affect the physical attributes of the water and sediment, including rate of soil accretion and the degree of anoxia in creek water, groundwater, and interstitial water. The presence of the trees and other biogenic structures act is a feedback loop, providing friction to slow movement of water and sediment. Even the forest canopy affects other ecosystem properties: by depressing wind and offering shade, the canopy assists in altering evapotranspiration and movement of tidal water and gases. As we shall see in the remaining chapters, these characteristics greatly affect life and energy flow in mangrove forests and waterways.

Chapter 4
Life in Tidal Waters

4.1 Introduction

Plankton communities living in mangrove waters are well adapted to the physics of water motion. The chemical properties of tidal waters are also affected by hydrology, with the relationship among plankton and nekton, water physics, and chemistry a nexus of consequential and inconsequential links. A consequence of secondary flows induced by strong bottom friction and complex topography is the aggregation of floating mangrove debris which attracts a menagerie of bacteria, protists, zooplankton, and fish. Likewise, tree stems, roots, and fallen timber—and their epiphytes—are attractive to many organisms entering the forest on the flood tide.

Mangrove plankton and nekton live in a murky milieu; the water is stained a brownish-green hue by the presence of polyphenolic compounds leached from litter and by suspended fine silt and clay particles. Mangrove waterways and channels are a harsh environment, with rapid currents or stagnant pools of water and variable temperatures, salinities, suspended solid loads, oxygen, and pH. Whether a source of food, shade, or refugia, mangrove forests are an important habitat for coastal organisms that either float or swim on the ebb and flood of the tide.

4.2 Physicochemical and Biochemical Attributes

Unlike the oft-described translucent waters of the tropics, tidal waters flowing into and out of mangrove waterways are relatively opaque. The low clarity of mangrove waters is the result of many interlinked processes, such as scouring and transport of fine particles from the forest and creek bed and in the flood tide from adjacent coastal waters. Rich plankton life and their bodily functions also contribute to literally muddying the waters.

Tidal waters entering and leaving mangrove waterways vary greatly in their physicochemical characteristics (see Chapter 3). Salinities can range from fresh-water to hypersaline depending on the extent of flushing and rainfall patterns. In many tropical estuaries, there is complete flushing because of monsoonal rainfall

D.M. Alongi, *The Energetics of Mangrove Forests*,

while in others, or often in the same estuary, water is trapped in the upper reaches for extended periods during the dry season. An inverse estuary can form if long residence times are accompanied by high evaporation rates at the mouth of the estuary (Wolanski 2007). Thus, mangrove biota must be either tolerant of a wide range of salinities or able to migrate into open coastal waters.

Dissolved oxygen and dissolved nutrient concentrations also vary greatly within and among mangrove estuaries, with values depending on geomorphology, salinity, catchment size, tidal range, precipitation patterns, biological activity, and the degree of vertical and horizontal mixing of creek waters (Alongi et al. 1992). One often-recurring pattern along mangrove waterways is a link between pH, oxygen and DOM concentrations, especially polyphenolic compounds. Moving upstream, especially when rainfall has been limited, it is common to observe a decline in both pH and oxygen concentrations but an increase in DOM. Boto and Bunt (1981) originally encountered this phenomenon in an Australian tidal creek, and attributed this pattern to oxidation of polyphenolics which would lower pH and oxygen levels. An alternative explanation is that pelagic respiration increases upstream when waters are stagnant, or experiencing long flushing times, enabling microbial consortia to concentrate and bloom, creating more particulate organic matter. Respiration would lower oxygen concentrations and the subsequent production of carbonic acid from CO_2 release would lower the pH.

Recent work on the chemistry of mangrove tidal waters has focused on characterizing and identifying the sources of DOC and DON (Hernes et al. 2001; Scully et al. 2004; Maie et al. 2008). The imperative driving such work is that DOM plays a pivotal role in microbial growth, the complexation of metals, nutrient availability, the extent of light penetration, and organic export (see Section 6.2). The bulk of the DOC is sourced from the leaching of mangrove leaves which contain high concentrations of tannins and other phenolic compounds to ward off herbivores (Hernes et al. 2001). Some individual compounds undergo rapid transformation by microbes, photooxidation, and abiotic processes, and may have an impact on the subsequent bioavailability of DOM. Polyphenolic compounds react and bind to sugars, proteins, lipids, and enzymes, making these labile substances less available for biological use (Maie et al. 2008). These processes facilitate formation of highly refractory, highly-colored, high molecular weight polymer-like substances derived from mangroves (Scully et al. 2004). This material is derived not only from leaching of mangrove leaves but also from mangrove porewater, probably from lateral transport of drainage from the forest floor at ebb tide (Tremblay et al. 2007). This means that much of the DOM in mangrove tidal waters is unaltered via photolysis, but that decomposition in the forest floor has selectively removed specific biodegradable molecules, leaving lignin as a significant DOM component to be exported by tides.

The complexation of these recalcitrant molecules with more labile compounds can affect nutrient recycling, as they often do in terrestrial forests (Krause et al. 2003). The mechanisms for this phenomenon have been investigated by Maie et al. (2008) using leaves, sediment, and natural waters from a *Rhizophora mangle* stand in Florida. In a series of experiments, Maie et al. (2008) examined the fate of leaf-derived tannins and found that (1) proteins were released under light exposed conditions; (2) the tannins underwent structural changes; (3) a large portion of tannins precipitated in seawater and bound to sediments; (4) the chemical half-life of tannins was <1 day; and (5) DON coprecipitated with tannins in natural tidal waters. Based on these findings, Maie et al. (2008) concluded that tannins assist in conserving nitrogen (as DON) and buffer its cycling by preventing rapid DON export and loss via rapid microbial breakdown; protein release from tannin-protein complexes is the rate-limited step in this process.

4.3 Loops, Chains, and Hubs in the Microbial Machinery

Prior to the mid-1970s there was little understanding of the energetic role of microbes in the sea. Through an explosive increase in improvements in microbial techniques since the mid-1980s, we now know that a large proportion of DOM and energy flow is funneled through highly diverse, actively growing, assemblages of Archaea, bacteria, ciliates, flagellates, amoebae, and viruses – many of mixed trophic states – and subsequently transferred to higher consumers via a chain of small protistan grazers in what essentially is a "microbial loop" (Fenchel 2008; Strom 2008). Originally, the concept of the microbial loop referred to the uptake by bacteria of small molecular weight DOM derived from plankton, even mucus produced by jellyfish. This material was thought to be shunted back through the food web via bacteriovores, such as ciliates, flagellates, amoebae, and other phagotrophs. The concept of the microbial loop has evolved (Landry 2001) to recognize the functional complexity of the intricate microbial machinery of planktonic food webs (Fenchel 2008; Strom 2008) as more of a "microbial hub" than loop (Legendre and Rivkin 2008) in that the microbes are tightly integrated with the classical food web (Fig. 4.1). Production and consumption within the hub is largely passed to higher consumers via multiple trophic transfers or lost via remineralization within the euphotic zone.

The study of microbial consortia in mangrove waters has lagged behind the study of forest life, including within the forest floor. This is puzzling given the large number of studies on the microbial hub in other aquatic environments. Also lacking is knowledge of species diversity among the various microbial groups.

Many more studies have estimated bacterial numbers in mangrove waters than their growth rates or productivity (Table 4.1), but it was not until the early 1990s that the first studies adequately investigated bacterioplankton dynamics. Indeed, in their review chapter on plankton, Robertson and Blaber (1992) were able to identify only four studies that enumerated water-column bacteria using a direct counting method. Since that time, more investigations have been conducted on aspects of the microbial loop, particularly spatial and temporal changes in abundance of bacteria and bacterioplankton productivity (Table 4.1).

Abundance of mangrove bacterioplankton ranges between 10^5–10^6 cells ml^{-1} (Table 4.1), which is within the range of cell numbers found in other tropical and temperate coastal waters (Ducklow and Shiah 1993). That bacterioplankton densities are usually within such a comparatively narrow range has been attributed to grazing pressure as this cell density range may represent a threshold below

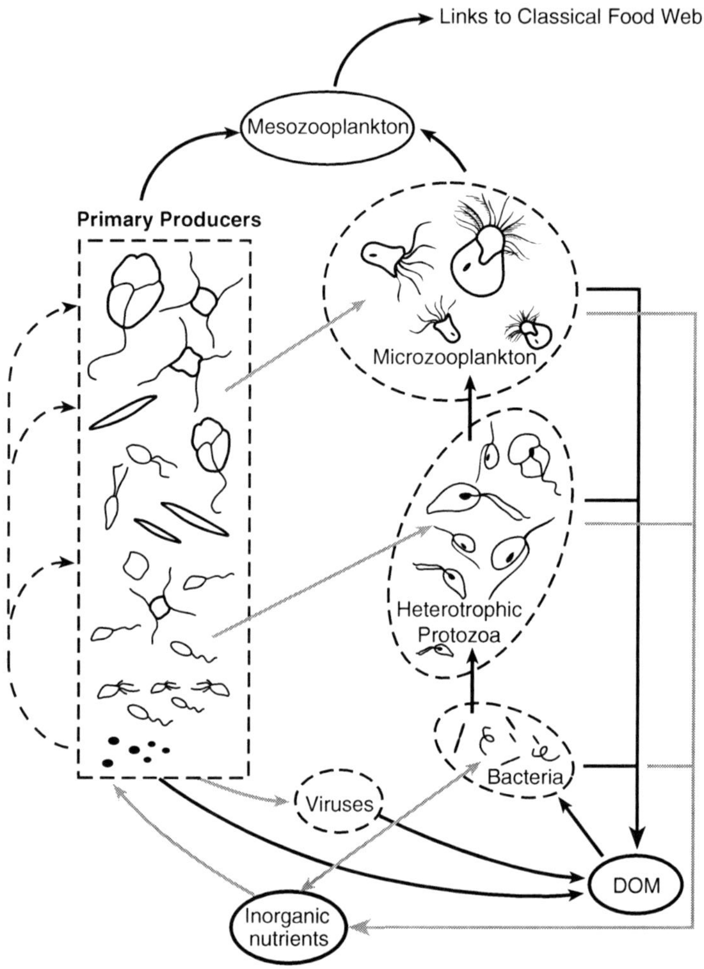

Fig. 4.1 A simplified visualization of the microbial consortia and their interrelationships in the microbial hub (Adapted from Landry 2001)

which capture by bacteriovores becomes functionally and energetically inefficient (Thelaus et al. 2008). Cell densities can vary with changes in tidal flow, proximity to the forest floor, concentrations of suspended particles, DOM and dissolved inorganic nutrients, the onset of monsoons, temperature, predation, and phytoplankton productivity, to name a few regulatory factors. For example, in the Gambia River, West Africa, most bacteria are free in the water-column, but densities of bacteria attached to particles vary in synchrony with the tide (Healey et al. 1988). When tidal waters are at maximum rates of ebb flow, attached bacterial densities also peak, especially in the wet season when suspended matter is at maximum concentration; at slack tide, numbers decline as particle-bacteria complexes settle. Substrate availability is a key factor controlling bacterial numbers in this estuary. This appears to be true for other mangrove water bodies, as eutrophic conditions with accompanying

Table 4.1 Estimates of bacterioplankton abundance (cells ml^{-1}) and productivity (μg C l^{-1} h^{-1}) in mangrove waters

Location	Abundance	Production	Production method	Reference
Gambia River, West Africa	1–2 × 10^{6}			Healy et al. (1988)
Biétri Bay, Ivory Coast	5.9–32.3 × 10^{6}	11–91	[^{3}H]-thymidine	Torréton et al. (1989)
Dona Paula estuary, India	1–6 × 10^{5}	0.7–3.9	[^{14}C]-glucose	Gomes et al. (1991)
Queensland, Australia	0.9–3.3 × 10^{6}			Robertson and Blaber (1992)
Fly River, Papua New Guinea	0.4–2.1 × 10^{4}	0.1–5.3	[^{3}H]-thymidine	Robertson et al. (1992b)
Indus River, Pakistan	1–4 × 10^{6}	2.1–12.5	[^{14}C]-leucine	Bano et al. (1997)
Sawi Bay, Thailand	0.9–9.2 × 10^{6}	1.4–16.5	[^{3}H]-thymidine	Ayukai and Alongi (2000)
Matang estuary, Malaysia	1.0–79.5 × 10^{6}	0.4–16.7	[^{3}H]-thymidine	Alongi et al. (2003b)
Guanabara Bay, Brazil	1.0–6.9 × 10^{6}	2.0–7.4	[^{3}H]-leucine	Andrade et al. (2003)
Ciénaga Grande, Colombia	6.5–90.5 × 10^{6}			Gocke et al. (2004)
Port Klang, Malaysia	2.5–9.8 × 10^{6}	3.9–6.1	Dilution culture	Lee and Bong (2006)
Cape Rachado, Malaysia	2.7–28.4 × 10^{6}	1.2–3.3	Dilution culture	Lee and Bong (2007)
St. Kilda, South Australia	6.9–7.5 × 10^{7}			Seymour et al. (2007)
Cananéia-Iguape, Brazil	0.4–2.3 × 10^{6}	1.5–22.0	[^{3}H]-thymidine	Barrera-Alba et al. (2008)

high suspended solid loads and water-column anoxia co-occur with high densities of bacteria (Torréton et al. 1989; Gocke et al. 2004).

Rates of bacterial productivity in mangrove waters range from 0.1–22.0 μg C l^{-1} h^{-1} in relatively pristine estuaries, and from 10–91 μg C l^{-1} h^{-1} in eutrophic waterways (Table 4.1), greater than the range of rates measured in salt marsh waters (4–5 μg C l^{-1} h^{-1}, Ducklow and Shiah 1993). Specific growth rates and rates of bacterioplankton productivity in most coastal waters often mirror rates of phytoplankton production, reflecting the first link in the microbial hub in which exudates from phytoplankton cells and cell contents released during 'sloppy feeding' by consumers are captured by bacteria to fuel growth. Unfortunately, the few studies available (Table 4.1) encompass mostly 'snapshot' views of bacterial dynamics in mangrove waters, addressing only limited areas and seasons. Further, only six studies measured production using the most viable techniques such as uptake of radiolabelled thymidine or leucine.

The importance of phytoplankton activity for fueling bacterioplankton growth is seen in a comparison of the ratios of bacterial to phytoplankton productivity

(Table 4.2). The BP/PP ratios are high (mean = 106%) compared with those of other estuaries for which Ducklow and Shiah (1993) calculated an average of 17%. This comparision can be misleading as the studies considered by Ducklow and Shiah (1993) were limited to those where seasonal cycles were adequately examined, unlike most of the studies in Table 4.2. These high ratios, even assuming a very high bacterial conversion efficiency, imply that other sources of organic matter, such as mangrove detritus, are being used to sustain bacterioplankton growth. The reasons for the high BP/PP ratios in mangrove waters might reflect comparatively low rates of phytoplankton production (see next section) or the efficiency with which mangrove bacterioplankton convert organic matter into biomass. In creek waters in Thailand, bacterioplankton and other microheterotrophs are supported by DOC (50–100%) and DON (40–90%) excreted by primary producers (Kristensen and Suraswadi 2002). In Dona Paula estuary, India, Gomes et al. (1991) found that 80% of phytoplankton exudates are removed by heterotrophic bacteria within 2–4 h. In similarly eutrophic waters at Port Klang, Malaysia, bacterial growth efficiencies of 6–22% were reported (Lee and Bong 2006). Bacterial conversion efficiencies for pristine mangrove waters are unknown, but even assuming a median efficiency of 50% (most recent empirical studies indicate lower values), carbon fixed by phytoplankton is insufficient to meet bacterial demand. The high rates of bacterioplankton production in relation to available fixed carbon can only be reconciled if (1) there is intense recycling within the microbial machinery and/or (2) organic matter from allothonous sources, such as mangrove detritus and benthic algae, supplement bacterial requirements.

The intensity of trophic interactions and recycling within the microbial hub in mangrove waters is virtually unknown (McKinnon et al. 2002b; Lee and Bong 2006, 2007). There is some evidence that viruses can exceed the abundance of bacteria (10^7 cells ml^{-1}) by an order of magnitude (10^8 cells ml^{-1}) (Seymour et al. 2007), but the functional significance of viruses in mangrove waters is unknown. Heterotrophic microflagellates range in abundance from 10^3–10^4 cells ml^{-1} (Bano et al. 1997), a range typical in other estuaries, and other bacterial grazers, such as naked amoebae, have been found in mangrove waters attached to mostly to flocs, ranging in abundance from 16–397 ml^{-1} (Rogerson and Gwaltney 2000; Rogerson et al. 2003).

Table 4.2 Ratio of volumetric rates of bacterial production to phytoplankton (BP/PP) production in mangrove estuaries. Studies only using thymidine or leucine to measure bacterial production are listed

Location	BP/PP ratio (%)	Reference
Biétri Bay, Ivory Coast	86%	Torréton et al. (1989)
Fly River, Papua New Guinea	84%	Robertson et al. (1992b)
Indus River delta, Pakistan	200%	Bano et al. (1997)
Sawi Bay, Thailand	55%	Ayukai and Alongi (2000)
North Queensland, Australia	53%	McKinnon et al. (2002a,b)
Matang estuary, Malaysia	63%	Alongi et al. (2003b)
Cananéia-Iguape, Brazil	198%	Barrera-Alba et al. (2008)

Densities of amoebae can exceed those of ciliates and other protists in the water-column (Buskey et al. 2004). Again, their trophic importance has not been demonstrated, but can be inferred, as amoebae are known to be voracious consumers of bacteria attached to particles (Landry 2001).

Grazing experiments conducted using planktonic microbes from mangrove waters point to high consumption rates of bacterioplankton (McKinnon et al. 2002b; Lee and Bong 2006, 2007). In Port Klang waters, consumption rates range from $5.5–26.9 \times 10^4$ cells ml^{-1} or 18–72 cells $protist^{-1}$ h^{-1}, equivalent to 20% of bacterial production. In nearby mangroves of Cape Rachado, Lee and Bong (2007), again using size fractionation–incubation methods, measured lower consumption rates of $0.5–5.7 \times 10^4$ cells ml^{-1} or 1–7 cells $protist^{-1}$ h^{-1}, indicating that 22% of bacterial carbon was passed on to protists. The remaining bacterial carbon is presumably grazed by other plankton and recycled with the microbial hub. There are also conditions in which grazing rates can exceed rates of bacterioplankton productivity, as McKinnon et al. (2002b) found in an Australian mangrove creek polluted by aquaculture effluent. The role of microbes in the balance between autotrophy and heterotrophy must be evaluated, but first we must understand the dynamics of phytoplankton.

4.4 Phytoplankton Dynamics

Mangrove phytoplankton communities are thought to be of low species diversity, owing to the inhibitory effects of the high concentrations of tannins and other polyphenolics. Few studies have attempted to identify phytoplankton taxa to the genus or species level beyond some extensive work in mangrove waters of India (Robertson and Blaber 1992). Species composition and the abundance of size fractions varies greatly among locations, but in pristine estuaries the nanoplankton and micro-plankton communities usually dominate; larger net phytoplankton, such as pennate and centric diatoms, are common but generally do not constitute the largest size fraction either in numbers or biomass, except in polluted waterways. Two notable exceptions are the Indus River delta, where in slightly eutrophic waters the phytoplankton community is comprised mostly of centric diatoms (Harrison et al. 1997) and the Achara River estuary in western India, where nanoplankton are dominant in the dry season but diatoms bloom in the monsoon and post-monsoon seasons, stimulated by heavy rains that flush elevated concentrations of nutrients into the estuary (Dham et al. 2005).

Total biomass and productivity of phytoplankton, like their species composition, vary greatly among regions. It does seem certain that biomass and primary production are greater in polluted or stagnant waterways, such as many in India and Brazil, or in creeks where flushing times are very long (Table 4.3). Rates of phytoplankton production also depend on the method used, as the ^{14}C and O_2 methods do not always agree; both have their shortcomings. For pristine mangrove estuaries, a 'typical' range of phytoplankton production is on the order of

Table 4.3 Phytoplankton biomass (as chlorophyll a = μg l^{-1}) and net primary production (NPP = mg C m^{-3} day^{-1}) in some mangrove estuaries

Country	Chlorophyll *a*	NPP	NPP method	Reference
Gambia	0.3–8.2	1–445	^{14}C	Healey et al. (1988)
Mauritana	0.2–3.6	215–580	O_2	Sevrin-Reyssac (1980)
Guadeloupe	10–60	8–1,700	^{14}C	Ricard (1984)
Brazil	1.1–19.3	110–500	^{14}C	Kutner (1975), Barrera-Alba et al. (2008)
India	4.4–39.8	60–662	O_2	Gomes et al. (1991)
Malaysia	0.5–43.2	22–755	^{14}C	Lee et al. (1984), Alongi et al. (2003b)
New Guinea	0.3–5.1	22–693	^{14}C	Robertson et al. (1992b)
Ivory Coast	18–94	526	O_2	Torréton et al. (1989)
Thailand	2–12	200–600	^{14}C	Ayukai and Alongi (2000)
Pakistan	1–8	20–195	^{14}C	Bano et al. (1997), Harrison et al. (1997)
Australia	0.3–3.5	25–212	^{14}C	Ayukai and Miller (1998a, b), McKinnon et al. (2002a, b, 2006)
Kenya	0.1–3.4	19–93	^{14}C	Hemminga (1995), Kitheka et al. (1996)

20–500 mg C m^{-3} day^{-1} and phytoplankton biomass is usually within the range of <1–3.5 μg chl *a* l^{-1} within a euphotic zone often <1 m. While the standing stocks of phytoplankton biomass are lower, the rates of phytoplankton production in mangrove waters are, on average, at the high end of the range of values measured in salt marshes and at the mid-range for other temperate estuaries (cf. Chapter 3, Alongi 1998). These habitat differences likely reflect greater temperatures and higher rates of production of emergent vegetation that may pass on greater levels of labile organic material.

Phytoplankton production and species composition in mangrove estuaries are regulated by a constellation of factors: light, temperature, nutrients, herbivory, water residence time, tidal energy, salinity, and lateral trapping and mixing of water. Light is the primary factor controlling phytoplankton photosynthesis in estuaries and coastal water; given the high turbidity and low light penetration in most mangrove waters (the euphotic zone is usually <1 m) mangrove phytoplankton production is undoubtedly light-limited. In macrotidal estuaries of northern Australia, there is a high degree of turbulence and turnover of the water-column, sustaining a very shallow euphotic zone (Burford et al. 2008). Suspended solid concentrations are high even in the dry season (≈6 mg l^{-1}) and secchi depth typically varies from 1–2 m.

Seasonality of phytoplankton production in the tropics and subtropics depends on distance from the equator and regional conditions, such as rainfall patterns, rather than on changes in water temperature. In the Indus River delta, there are no seasonal patterns (Harrison et al. 1997), but in Darwin Harbour, a macrotidal estuary in Australia, phytoplankton production is greater in the wet season (2.2 g C m^{-2} day^{-1}) than in the dry season (1.0 g C m^{-2} day^{-1}) with no significant difference in levels of chlorophyll *a* (McKinnon et al. 2006). Wet season-dry season differences in phytoplankton production can also be found in lagoons and estuaries in Mexico, southeast India, and Kenya, where rates of net primary production often peak at the onset of the rainy season (Kitheka et al. 1996; Rivera-Monroy et al. 1998; Purvaja and Ramesh 2000).

Phytoplankton production in mangrove estuaries can be limited by other factors in addition to light. For instance, in Darwin Harbour, there is some evidence for nitrogen limitation (Burford et al. 2008). Nitrogen limitation can also occur in mangrove estuaries in the dry tropics as in Exmouth Gulf on the arid tropical coast of Western Australia, where phytoplankton production rates average <25 mg C m^{-3} day^{-1} (Table 4.3). An analysis of mangrove plankton studies conducted by the Australian Institute of Marine Science where phytoplankton production, suspended solids, and nutrients were measured simultaneously, shows that phytoplankton production in Australian mangrove waters correlates only with dissolved inorganic phosphate (Fig. 4.2) implying phosphorus rather than nitrogen limitation. The higher concentrations of DIN relative to DIP in most mangrove estuaries – frequently above the Redfield ratio (Alongi et al. 1992) – supports this supposition. Other mangrove studies point to P limitation for phytoplankton (Mohammed and Johnstone 1995; Harrison et al. 1997; Holmer et al. 2001; Kristensen and Suraswadi 2002). Depending on local conditions, mangrove phytoplankton can be limited by phosphorus or nitrogen or both, in addition to light. In dilution experiments in Exmouth Gulf, phytoplankton growth and grazing were stimulated in treatments receiving additional nitrate (Ayukai and Miller 1998a). The switching between N and P limitation in mangroves is hardly surprising considering that mangroves are at the interface between freshwater where P can be limiting and marine waters where N is often limiting.

Phytoplankton cells are capable of taking up and assimilating various dissolved nitrogen compounds, with preferences depending on size class and relative availability of nutrient species. In the mangrove waters of western India, nitrate is the major fraction assimilated (72%), followed by NH_4^+ (16%), nitrite (6%), and urea (6%), although there was seasonal switching of preferences (Dham et al. 2002).

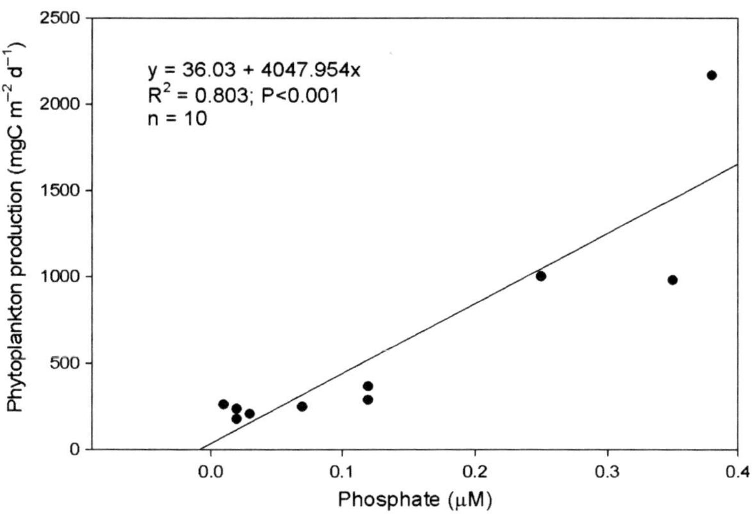

Fig. 4.2 The relationship between phytoplankton production and phosphate concentrations in mangrove estuaries of tropical Australia (Data from Ayukai and Miller 1998a, b; McKinnon et al. 2002b, 2006; and D.M. Alongi unpublished, 2006, 2007)

In the post-monsoon, nitrite and nitrate are preferred, and in the post-monsoon season, NH_4^+ is preferentially used. This switching was attributed to higher relative abundance and strong allochthonous supply of nitrite and nitrate compared to ammonium which was relatively more abundant after the monsoon because of higher regeneration rates. Size-fractionation experiments in these waters show that pennate diatoms are responsible for >90% of nitrate uptake whereas cyanobacteria account for >80% of the NH_4^+ uptake (Dham et al. 2005). Ammonium regeneration and nitrification in the water column provide ≈20% more N than assimilated by the phytoplankton. Dham et al. (2002) suggest that this "excess" nitrogen is taken up by the forest and benthic algae. If true, there may indeed be a direct biogeochemical link between the phytoplankton and the forest.

4.5 Are Mangrove Waters Net Heterotrophic or Autotrophic?

An understanding of the metabolic balance between production of carbon fixed by primary producers (gross primary production, P_G) and carbon respired by all organisms (community respiration, R) is important to the global carbon cycle and for an assessment of the state of an ecosystem. When the ratio of gross primary production to community respiration (P_G/R) is >1 more carbon is being fixed than is lost via respiration, so the ecosystem is considered net autotrophic. When $P_G/R = 1$ the ecosystem is in overall balance. When $P_G/R < 1$, the system is losing more carbon than is being fixed, and is considered net heterotrophic. The presumption in the latter case is that allochthonous materials are imported to sustain carbon balance. A few studies have examined gross primary production and pelagic respiration in mangrove waterways (Fig. 4.3), so we may ask the question: Are mangrove waters net heterotrophic or autotrophic, and under what conditions?

A global analysis of carbon metabolism in coastal ecosystems concluded that estuarine waters are net heterotrophic, with a mean P_G/R of 0.8 and a standard deviation of 0.05 (Gattuso et al. 1998). A similar analysis (Alongi 1998) concluded that estuaries are in a state of overall balance ($P_G/R = 1.0$), but at the time of these analyses data for mangrove waters were insufficient, especially for pelagic respiration. Most of the data points (Fig 4.3) lie above the $P_G/R = 1$ line indicating net autotrophy, with GPP exceeding pelagic respiration rates, in most mangrove waters. The mean P_G/R is 1.8 with a standard deviation of 0.3, substantially higher than the P_G/R's of 0.8 and 1.0 calculated for other estuaries (Gattuso et al. 1998; Alongi 1988). The values <1 come mostly from the study of a creek on Phuket Island, Thailand, which was sampled only in the dry season (Kristensen and Suraswadi 2002). This creek is shallow (<4 m) and surrounded by shrimp ponds that drain into smaller channels along the waterway that may have played a role in shifting the creek to a state of net heterotrophy. The data set is limited and likely to mask the influences of factors such as water depth, extent of freshwater input, suspended solid concentrations, and tidal range.

Within a particular water body, the metabolic state is likely to change with season or over longer periods of time. A seasonal shift between net autotrophy

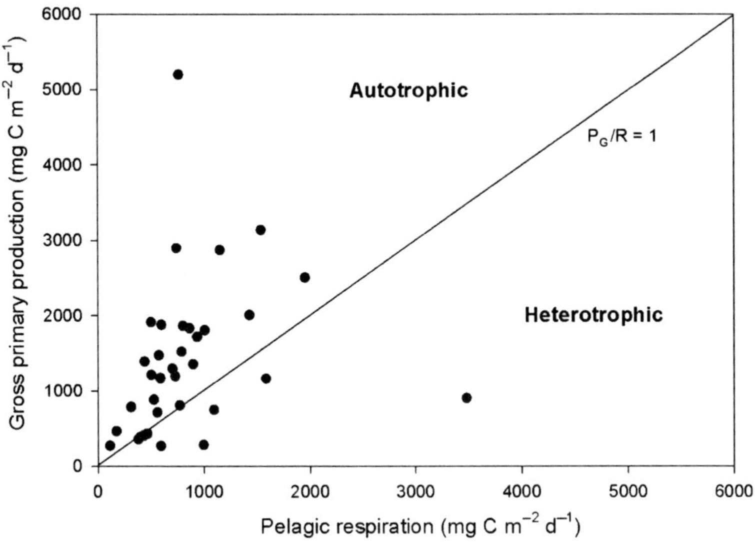

Fig. 4.3 The relationship between rates of gross primary production and community respiration in mangrove waters. The solid line represents $P_G/R = 1$ (Data from Ayukai and Miller 1998a, b; Ayukai and Alongi 2000; Kristensen and Suraswadi 2002; McKinnon et al. 2002b, 2006; Alongi et al. 2003; Ram et al. 2003; Gocke et al. 2004; and Burford et al. 2008)

and heterotrophy was discovered in the Mandovi and Zuari estuaries in southwestern India (Fig. 4.4). In the pre- and post-monsoon months when conditions are quiescent and water clarity is maximal, waters are net autotrophic (Fig. 4.4). During the monsoon season, when light levels and salinity are at a minimum and the waters are at maximum turbidity (from sediment resuspension and peak loads of excess organic matter from upstream), rates of pelagic respiration are enhanced and phytoplankton productivity is suppressed, leading to a state of net heterotrophy (Ram et al. 2003). On an annual basis, these two estuaries are net autotrophic as the non-monsoon seasons are longer than the monsoonal period.

Rates of pelagic respiration in mangrove creeks and rivers average 846.9 mg C m^{-2} day^{-1} and range from 0.1–3.5 g C m^{-2} day^{-1} (Fig. 4.3), less than the mean rate of 1,368 mg C m^{-2} day^{-1} in temperate and sub-tropical estuarine waters (Hopkinson and Smith 2005), although there is significant overlap in ranges, and water depth is a confounding factor. Also, a variety of methods have been used to measure pelagic metabolism, limiting comparisons among ecosystems. Temperature, nutrients, grazers, availability of organic substrates, and a host of other factors regulate pelagic respiration rates in temperate waters (Hopkinson and Smith 2005) and likely do so in mangroves.

An analysis of factors measured in conjunction with respiration reveals only one significant correlation (Fig. 4.5) of respiration with chlorophyll *a* standing stocks. The other factors tested were temperature, bacterial numbers, water depth, and salinity; too few data are available to meaningfully correlate pelagic respiration with bacterioplankton production. The significant relationship between standing stocks

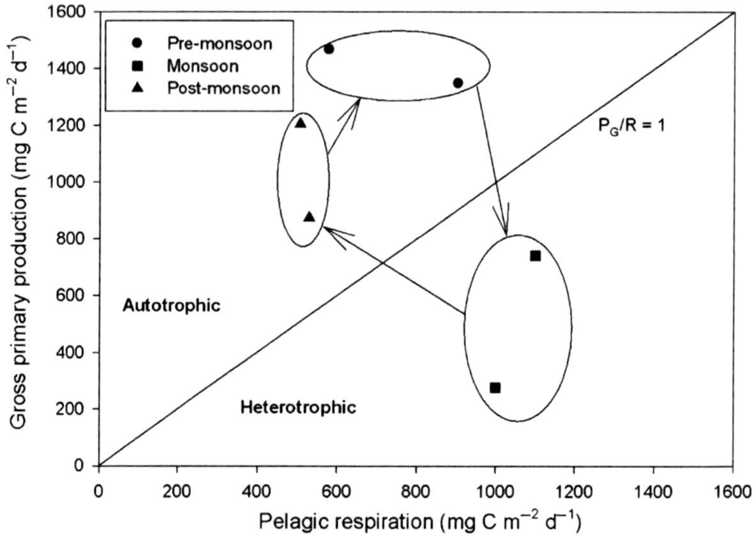

Fig. 4.4 Seasonal shift in the metabolic state of plankton communities in the waters of the Mandovi and Zuari estuaries, southwestern India (Data from Ram et al. 2003)

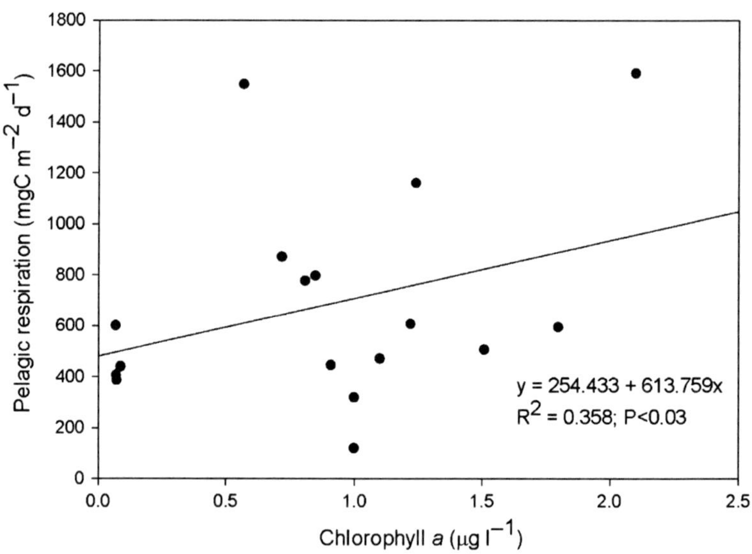

Fig. 4.5 The relationship between pelagic respiration and chlorophyll *a* in mangrove estuarine waters (Data are from Ayukai and Miller 1998a, b; Ayukai and Alongi 2000; Kristensen and Suraswadi 2002; Alongi et al. 2003b; McKinnon et al. 2006)

of chlorophyll *a* and respiration implies linkage between microheterotrophs and phytoplankton, but may also indicate that phytoplankton contribute significantly to total respiration. Variation not explained by chlorophyll *a* is presumably due partly to the influence of allochthonous organic inputs, or on factors not measured in these studies. Bacteria are responsible for most CO_2 respired in the water column, but other heterotrophic and autotrophic organisms, including yeasts, ciliates, amoebae, flagellates, and invertebrates of small size, can make a very variable contribution to total community respiration.

A maximum value for bacterial growth efficiency (BGE) can be calculated for the data, BGE = BP/(BP + BR), where BP = bacterial production and BR = bacterial respiration. Assuming that 50% of the mean respiration rate of 846.9 mg C m^{-2} day^{-1} is of bacterial origin (Hopkinson and Smith 2005) and using a mean value for bacterial production (BP) of 2,016 mg C m^{-2} day^{-1} (from the studies cited in Section 4.2), a minimum bacterial growth efficiency of 82% is calculated. BGE ranges from <30% to near 60% for growth on a variety of substrates (Cole 1999), so the high mangrove BGE suggests that either (1) bacterial production is overestimated, or (2) respiration is underestimated, or both. Aquatic ecosystems showing net autotrophy are characterized by either a low BP/NPP ratio or an especially high BGE (Cole 1999). Although highly unlikely to be as great as 82%, this fact argues for mangrove BGE's being at the high end of the 30–60% range.

The use of conversion factors in the methods used to measure bacterial production may play a big part in this discrepancy, underscoring the importance of determining factors such as bacterial carbon content empirically. Assimilation efficiencies of 6–40% have been measured for mangrove bacterioplankton (Gomes et al. 1991; Lee and Bong 2006), so we can assume that some of the rates of bacterial production are overestimates. The very high ratios of bacterial production to primary production (Table 4.2) further suggest that the bacterioplankton production values are overestimates. Obviously, more work needs to be done to accurately determine the contribution of bacterioplankton communities in mangrove estuaries.

4.6 Zooplankton

Zooplankton assemblages are the crucial link between members of the microbial hub (to which some of the smallest zooplankton belong, Fig. 4.1) and larger consumers, particularly penaeid shrimps and zooplanktivorous fish. Zooplankton are operationally grouped into three size classes: (1) the microzooplankton (20–200 μm) which include a menagerie of protists, rotifers, veligers of various mollusks, and nauplii of a variety of crustaceans, such as barnacles and copepods; (2) the mesozooplankton (200 μm–2 mm), usually dominated by cyclopoid, calanoid, and harpacticoid copepods; and (3) the macrozooplankton (>2 mm) which in mangrove waters are composed mostly of jellyfish, ctenophores, chaetognaths, decapod larvae, mysid shrimp, and other gelatinous organisms; they feed mainly on mesozooplankton.

4.6.1 Factors Affecting Abundance, Composition, and Biomass

Most studies of mangrove zooplankton have focused on their community structure, distribution, and abundance in relation to tides, seasons, salinity, and various other physical factors. Only a précis is given here as many of the conclusions of Robertson and Blaber (1992) regarding mangrove zooplankton remain true. The main factor controlling species composition and abundance of mangrove zooplankton communities is the seasonal change in salinity. The onset of the monsoon season in most mangrove locations is the stimulus for clear shifts in community composition and abundance. There is a well-defined, year-round gradient in zooplankton community composition: (1) a stenohaline marine component that penetrates into the estuary mouth, (2) a euryhaline marine component that penetrates further up estuary, (3) a true estuarine component, and (4) a freshwater component.

Numbers and biomass of zooplankton vary widely among mangrove estuaries with densities ranging from 10^3 to 10^5 m^{-3} and biomass ranging from <1 to over 600 mg m^{-3}, but usually greater than the abundance and biomass of zooplankton communities in adjacent coastal waters (Robertson and Blaber 1992; Kathiresan and Bingham 2001). The most common pattern of abundance is peak densities in summer and lowest abundance and biomass during the monsoon season, mirroring changes in phytoplankton abundance and productivity. In Indian mangroves, tintinnid ciliates are the dominant group of microzooplankton and copepods are the dominant components of the mesozooplankton, as they are in most other mangrove estuaries.

An often conspicuous feature of mangrove zooplankton is the presence of species of the cyclopoid copepod family Oithonidae. A particularly fascinating species is *Diothonia oculata* which exhibits swarming behavior in light shafts among mangrove prop roots, formed in response to an endogenous rhythm and maintained in tidal currents up to 2 cm s^{-1} (Buskey et al. 1996). Oithonid copepods may have a selective advantage in mangrove waters where small prey such as pico- and nanoflagellates are abundant, suffering less predation than larger bodied copepods, and employing a number of strategies to maximize growth, reproduction, and survival (Duggan et al. 2008). In addition, oithonids can reproduce at temperatures lethal to other copepod species (Turner 2004) and have low metabolic rates that allow them to exploit low oxygen concentrations (Lampitt and Gamble 1982).

Vertical patterns of other zooplankton persist in the face of strong currents, but it is unclear if these patterns arise from active swimming or are the result of some physical process. Other copepod taxa characteristic of mangrove zooplankton communities are the calanoid genera *Acartia*, *Paracalanus*, *Parvocalanus*, and *Pseudodiaptomus*. Some species of these genera can to some degree utilize current shear to maintain horizontal position in an estuary (Kimmerer and McKinnon 1987). Many estuarine and marine zooplankton can migrate vertically and laterally to enhance survival, orienting themselves to a particular position in response to exogenous or endogenous stimuli. Naylor (2006) suggests that by differential exploitation of tidal or wind-driven currents, the orientation behavior of some zooplankton species constitutes a form of navigation.

Meroplankton (planktonic larval stages of benthic invertebrates) are transitory members of the zooplankton, but they can be seasonally dominant, especially with the onset of the breeding cycle for many crustaceans such as brachyuran crabs and decapods. On shorter time spans, positional changes can be induced by tides (Krumme and Liang 2004). In Brazilian estuaries, copepods maintain their position during weak neap tides, but during strong spring tides, some (but not all species) are carried out of the estuary. Some species tend to sink to the forest floor during flood tides to minimize horizontal export on the ebb tide. It is not known to what extent mangrove zooplankton exhibit vertical migratory behavior to be actively retained within the estuary.

4.6.2 Diets and Grazing Rates

Microzooplankton, dominated by protists, function within the microbial hub, exploiting aggregates of pico- and nanoplankton as well as other microzooplankton, such as ciliates, nauplii and veliger larvae, and detritus. Meso- and macrozooplankton feed on a wide array of foods, including microzooplankton and particulate organic matter, switching dietary preferences over time and space. Some zooplankton, such as cyclopoid copepods, undergo a dietary shift during their life stages, from mostly herbivorous nauplii to primarily carnivorous late copepodites and adults. The larger zooplankters play an important role as a trophic link to fish and as agents for structuring plankton communities. Also, their selective predation on the microherbivores implies that mesozooplankton grazing may be substantial, especially in oligotrophic waters (Calbet 2001; Buskey et al. 2003). However, in the open ocean and in coastal waters, microzooplankton tend to dominate mesozooplankton as consumers of phytoplankton, and their grazing often accounts for most phytoplankton production. Microzooplankton account for ≈75% of phytoplankton daily growth in subtropical and tropical waters, with respiration by protists averaging 35–43% of daily primary production – within the same order of magnitude as bacterial respiration (Calbet and Landry 2004; Putland and Iverson 2007). We can only presume that such is the case in mangrove waters.

Dietary studies of mangrove zooplankton have focused on the juvenile stages of penaeid shrimps (Newell et al. 1995), due to their obvious commercial importance, rather than on the more energetic micro- and mesozooplankton. Recent observations using stable isotope tracers confirm that juvenile shrimps ingest phytoplankton and mangrove—derived organic matter, but also show that the mangrove dietary signal rapidly decreases with distance from the shore (Bouillon et al. 2000; Kibirige et al. 2002). Suspended organic matter in mangrove waters is composed of a highly variable mix of terrestrial detritus and algae (including macroalgae) that changes seasonally and spatially (Cifuentes et al. 1996).

Other zooplankton feed discriminately, selecting phytoplankton cells among the suspended matter in creeks and waterways (Bouillon et al. 2000). In South Africa, Kibirige et al. (2002) found that mysids and copepods feed mainly on phytoplankton, ciliates, and flagellates, with no evidence of detritus consumption.

Laboratory feeding experiments confirm that calanoid and cyclopoid copopods feed on algae but also show that fecal material from grapsid crabs, when mixed with algae, result in greater assimilation rates (Werry and Lee 2005), forming a slender link between zooplankton and benthos.

Seasonality and other external factors (e.g., excess nutrient loads) play an important role in trophic relationships among plankton. In Cochin backwaters and in mangrove lagoonal waters in southeast India, grazing experiments show that microzooplankton consume 43% of the daily standing stock of phytoplankton during the dry months but during the monsoon, grazing impact on phytoplankton is masked by the dominance of freshwater flow and low residence times of very low densities of microzooplankton and mesozooplankton (Godhantaraman 2002; Jyothibabu et al. 2006). In Australian mangrove creeks receiving aquaculture wastes, McKinnon et al. (2002b) measured microzooplankton grazing rates in excess of 100% of phytoplankton and bacterioplankton production during periods of waste discharge; during non-discharge periods, grazing and growth rates were equivalent, returning to a state of trophic balance. In the dry tropics of Western Australia, ciliates, crustacean nauplii, and copepodites are the most important phytoplankton grazers, outpacing the rates of phytoplankton production by 30–90% (Ayukai and Miller 1998a). In such oligotrophic waters, other sources of organic matter are needed to balance ecosystem demand. A similarly high proportion of grazing mortality was measured in mangrove channels of Twin Cays in Belize (Buskey et al. 2004). In these tidal channels, the microzooplankton community was dominated by the swarming copepod *Dioithona oculata* followed by ciliates. Grazing experiments using mixed microzooplankton assemblages indicate that 60–90% of potential phytoplankton production was grazed. Ingestion rates of *D. oculata* were highest on ciliates and autotrophic dinoflagellates; the copepods were capable of grazing ≈10% of the standing stock of protists per day. Microzooplankton communities therefore exert tight control over phytoplankton production in mangrove waters, serving as an important trophic link between their microbial prey and their larger metazoan predators.

4.6.3 Secondary Production

Secondary production has been measured for only a few species of mangrove zooplankton, so the significance of zooplankton production to total energy flow in mangrove ecosystems is uncertain. Egg production rates of some dominant copepods were first measured by McKinnon and Klumpp (1998) and McKinnon and Ayukai (1996) in Australian mangroves. They measured egg production rates in the range of <1–51 eggs female^{-1} day^{-1}, which is relatively high compared with rates measured for copepods in other marine environments. Production rates are lower in dry tropical systems than in the wet tropics. Statistical models suggest that factors other than temperature (e.g., food availability) regulate egg production. A similar study of egg production of the calanoid copepod *Acartia lilljeborgi* in the Cananéia Lagoon of Brazil measured high rates of egg production (13.8–66.8

eggs female^{-1} day^{-1}) and found a significant relationship with temperature and standing stocks of chlorophyll *a* (Ara 2001). Ara (2002) later measured secondary production of the calanoid copepod *Temora turbinate* in the same lagoon and calculated egg production rates of 0.0002–1.115 mg C m^{-3} day^{-1} with P/B ratios varying from 0.17–0.45 day^{-1}.

Secondary production of zooplankton may be related to temperature and food availability, but the influence of these two factors may be timed to the onset of the monsoon. This idea is corroborated by studies in Indian estuaries. In the Pichavaram mangroves near Parangipettai, southeast India, secondary production estimates for the entire tintinnid community averaged 1.6 μg C l^{-1} day^{-1} during summer, 1.4 μg C l^{-1} day^{-1} in the post-monsoon season, 1.2 μg C l^{-1} day^{-1} in the pre-monsoon, but only 0.1 μg C l^{-1} day^{-1} during the monsoon, correlating inversely with changes in salinity, phytoplankton, and predator standing stocks (Godhantaraman 2002). Secondary production of zooplankton followed a similar pattern in a mangrove lagoon on the southwest coast of India (Nayar 2006). Clearly, further production studies are needed to assess the energetic importance of zooplankton in mangrove estuaries as the few studies to date suggest high production potential of mangrove zooplankton.

4.7 Nekton: Diets, Growth, and Trophic Links

4.7.1 Penaeid Shrimps

Excluding mud crabs, shrimps of the genera *Penaeus, Metapenaeus, Parapeneopsis*, and *Macrobrachium*, constitute the bulk of the commercially important crustacean landings from mangrove estuaries. It is therefore not surprising that great effort has focused on penaeid distribution, abundance, degree of habitat dependence, and catch per unit effort.

Higher densities of juvenile penaeids are found in most mangrove nursery areas than in adjacent habitats, such as sandflats (Rönnbäck et al. 2002). These juveniles eat diatoms, filamentous algae, epiphytic algae, mangrove detritus and attached microbes, foraminifera, hydrozoans, mollusks, nematodes, insects, fish larvae, and other crustaceans (Leh and Sasekumar 1984). Stable isotope studies confirm the importance of these foods, but found that shrimp feed on prey that derive their carbon and nitrogen from benthic algae and/or phytoplankton to a much greater extent than from mangrove detritus (Newell et al. 1995; Dittel et al. 1997). The relative contribution of algal versus mangrove carbon and nitrogen appears to be dependent on location. A detailed study of juvenile prawn nutrition in Malaysian mangroves and offshore waters (Chong et al. 2001) paints a more complex story of the origin of the organic matter consumed by *Penaeus merguiensis, Metapenaeus brevicorni, M. affinis, M. lysianassa, Parapenaeopsis sculptilis, P.coromandelica*, and *P. hardwickii*. A plot of the stable isotope results for the muscle tissue of these penaeids, mangrove leaves, plankton, and sediments from mangroves and offshore (Fig. 4.6) shows that the contribution of mangrove carbon to prawn tissues is high,

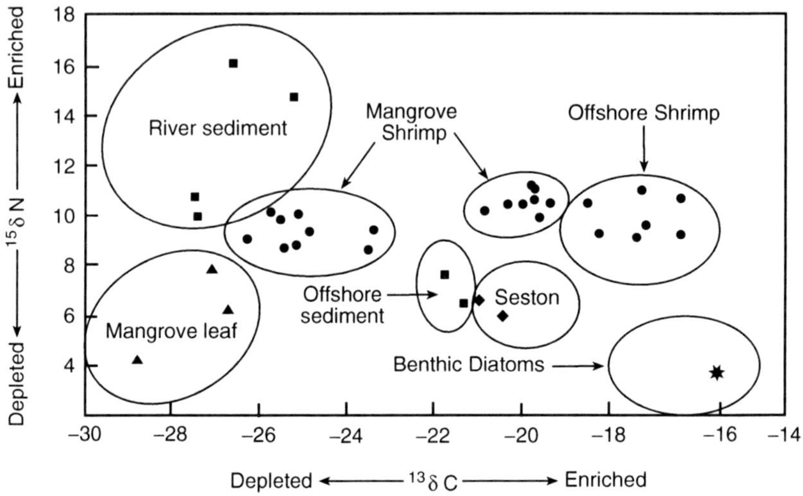

Fig. 4.6 The relationship between δ ^{13}C and δ ^{15}N for shrimp muscle tissues, mangrove leaves, and plankton and sediment from mangroves and adjacent offshore waters, west coast of peninsular Malaysia (Adapted from Chong et al. 2001)

but decreases offshore as the phytoplankton signal becomes progressively stronger. Shrimps from 2 km offshore still exhibit some mangrove dependency, but 7–10 km further offshore, the main diet is phytoplankton with some assimilation of benthic microalgae.

Shrimps function as mid-level and top omnivores, regulating the abundance of smaller plankton and nekton through predatory pressure, spending their postlarval and juvenile stages in the estuary until emigrating offshore during the wet season where they spawn (Robertson and Blaber 1992). While in the estuary, they feed non-selectively, tending to aggregate close to the water's edge near low tide. At flood tide, they move back up the estuary and enter smaller creeks and the forests.

The few estimates of annual production of penaeid shrimp per hectare of mangrove forest (Table 4.4) reflect catch per unit effort rather than true empirical measurements of secondary production. As fishing effort is not 100% efficient, the real production rates are undoubtedly higher. Nevertheless, the mean and ranges of values (Table 4.4) indicate that in most mangrove estuaries of Southeast Asia, rates of annual shrimp production are at the high end of the range of values (13–756 kg ha^{-1} $year^{-1}$) originally compiled by Turner (1977). This production represents a substantial economic value for each country, but it is often undervalued (Rönnbäck 1999).

Globally, annual landings of penaeid shrimp within and adjoining mangrove waters quadrupled from <100,000 t in 1950 to nearly 300,000 t in 1980, with a not-so-slow decline to ≈200,000 t by 2006 (Fig. 4.7). These data are worrying in that they clearly indicate overexploitation of penaeid shrimps in coastal mangrove waters. The picture is no better for mangrove fish.

Table 4.4 Annual shrimp production (kg ha^{-1} year^{-1}) in mangrove forests at various locations within Southeast Asia

Country	Shrimp production	Reference
Indonesia	16–165	Martosubroto and Naamin (1977)
Malaysia	515–700	Gedney et al (1982), Sasekumar and Chong (2005)
Philippines	130–150	Pauly and Ingles (1986)
Java, Indonesia	161	Naamin (1990)
Papua, Indonesia	18	Ruitenbeck (1994)
Perak, Malaysia	670	Singh et al. (1994)
Sumatra	274	Hambrey (1996)

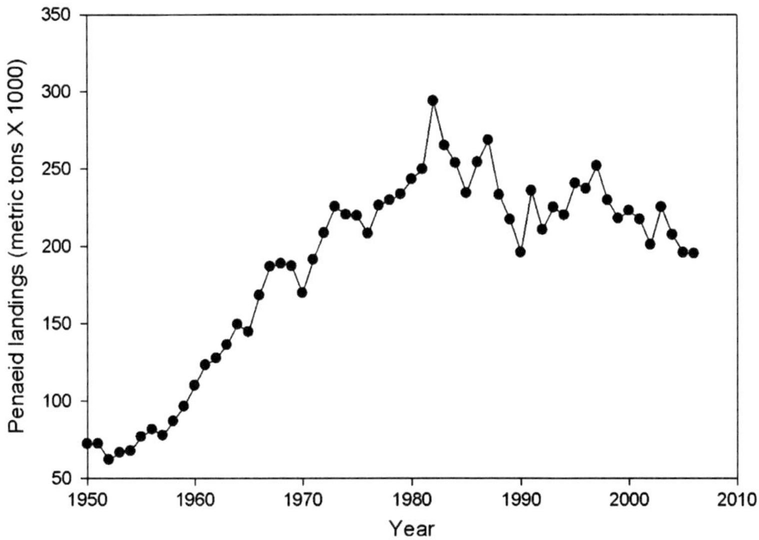

Fig. 4.7 Total annual landings of wild stock penaeid shrimps (metric tons × 1,000) from all countries with mangrove forests (Data from 2007 FAO Fishstat Plus software (http://www.fao.org/fishery/topic/16073))

4.7.2 *Fish*

Like the penaeids, most information on mangrove fish communities deals with species distribution and abundance patterns rather than on growth, mortality, and secondary production (Faunce and Serafy 2006). Species richness of both temporary and permanent residents is a function of microhabitat diversity (open water channels, mudflats), tidal amplitude, water clarity and depth, salinity, proximity to seagrass beds and coral reefs, and current patterns of the offshore environment. Numbers of fish species can range from <10 to nearly 200 in any particular estuary, with a tendency for more species in larger systems (Robertson and Blaber 1992). Densities and biomass of fish are difficult to compare owing to differences in equipment used and microhabitats sampled. The data complied by Robertson and Blaber (1992) indicate a density range of 1–161 fish m^{-2} and 0.4–29 g m^{-2} of biomass worldwide, greater than in temperate estuaries (Blaber 2002).

Fish inhabiting mangrove waters are grouped into five feeding guilds: herbivorous, iliophagus, zooplanktivorous, benthic invertebrate feeder, and piscivorous species. These feeding groups represent the results of a large number of studies devoted to analyses of fish gut contents (Robertson and Blaber 1992). Recent work using stable isotopes (e.g., Sheaves and Molony 2000; Kieckbusch et al. 2004; Lugendo et al. 2006) has confirmed the earlier work on fish diets, but has clarified gut contents previously labeled 'detritus', 'indistinguishable', or 'amorphous'. Sheaves and Molony (2000) confirm that several Australian fish species feed extensively on sesarmid crabs thereby 'short-circuiting' the traditional view of mangrove food webs. In subtropical lagoons in south Florida, much of the amorphous material in fish guts is mostly seagrass detritus rather than mangrove litter (Kieckbusch et al. 2004). Stable isotope analyses of juvenile fishes in an African bay lined with mangroves, mudflats, and seagrasses, found that crustaceans are the preferred food of zooplanktivorous and omnivorous fish, while fishes and algae are the preferred foods of piscivores and herbivores, respectively; there was a very limited signal for mangrove or seagrass detritus in the guts of most fish examined. Many species, however, shift their dietary preferences when advancing from larval to juvenile to adult stages.

Our knowledge of fish productivity in mangrove estuaries is limited. In Mexican lagoons with fringing mangroves, the secondary production of the mangrove resident *Fundulas parvipinnis* averages 0.32 g DW m^{-2} $year^{-1}$ (Pérez-España et al. 1998) and the transient *Mugil curema* at 2.7 g DW m^{-2} $year^{-1}$ (Warburton 1979). In the Bahamas, Valentine-Rose et al. (2007) estimated secondary production of the schoolmaster snapper *Lutjanus apodus* at 527 g WW m^{-2} $year^{-1}$. The most recent effort to accurately measure secondary production found that *Lutjanus griseus* production ranged from 1.34–2.66 g DW m^{-2} $year^{-1}$ (Faunce and Serafy 2008).

Although only four studies have actually measured secondary production of mangrove fish species, catch effort data is available from a number of locations as a proxy measure of fish productivity (Table 4.5). These estimates indicate catch

Table 4.5 Annual wild fish production (kg ha^{-1} $year^{-1}$) in mangrove forests at various locations worldwide

Country	Production (kg ha^{-1} $year^{-1}$)	Reference
Benin	560–860	
Philippines	90–1000	Janssen and Padilla (1999)
Ivory Coast	160	Marten and Polovina (1982)
Indonesia	140–630	Marten and Polovina (1982)
Colombia	120	Marten and Polovina (1982)
Madagascar	28–37	Marten and Polovina (1982)
El Salvador	17	Marten and Polovina (1982)
Mexico	732	Vega-Cendejas and Arreguin-Sánchez (2001)
India	613–1646	Kathiresan and Rajendran (2002)
Bangladesh	343–400	Islam and Haque (2004)

effort over a wide range (17–1,646 kg ha^{-1} $year^{-1}$); most estimates suggest landings mostly as a few hundred kilograms per hectare per year—somewhat greater than penaeid shrimp production (Table 4.4). The database is too scant to offer any indication of statistically significant differences among locations, but at least one study from southeast India (Kathiresan and Rajendran 2002) points to seasonal patterns in fish landings, with peak catch around the post-monsoon and summer months and lowest landings during the monsoon. It is unclear if these patterns reflect either temporal change in catch effort or seasonal patterns in recruitment of most fish species, or both.

The future for mangrove fish seems bleak. Plots of annual wild catch of the highly prized barramundi (or giant sea perch *Lates calcarifer*) and mangrove red snapper (*Lutjanus argentimaculatus*) in Southeast Asian waters (Fig. 4.8) show rapid growth in the catch of both species since the 1950s, but overexploitation since the end of the twentieth century. Data are difficult to obtain for other mangrove-associated fish species, but it is likely that most species caught in mangrove estuaries worldwide are either at or beyond maximum sustainable limits.

4.8 Is There a Link Between Mangroves and Fisheries Production?

The idea that mangroves are important nursery grounds for many edible species of estuarine and coastal fishes and crustaceans has been a key ecological paradigm since it was first articulated by Eric Heald and Bill Odum (Heald 1969; Odum and Heald 1972). Links between mangroves and fish were suggested earlier (Dakin 1938), and since much farther back in time indigenous people have been aware of possible links between mangroves and edible items. An exhaustive review of this subject was published recently (Manson et al. 2005a). This subject is very important as many key management decisions involving mangroves are based on this paradigm.

Three major hypotheses—not mutually exclusive—have been offered to explain the connection between mangroves and coastal fisheries; that mangroves offer (1) an abundant variety of foods, (2) a refuge from predation, most notably in the form of above-ground roots, fallen timber, litter, and shallow turbid water, and (3) shelter from physical disturbances. There is very little direct evidence of a clear reason for the relationship between fishery catch and mangroves; direct experimental tests have been rare (e.g., Cocheret de la Morinière et al. 2004) Most attempts to deduce such a link rely on correlation or regression analysis which is fraught with problems, such as auto-correlation. Also, as recently articulated (Sheaves 2005), most mangrove ecosystems are available to nekton for only part of the time, depending on the state of the tides, so many species migrate to and occupy alternative habitats, such as seagrass beds and even coral reefs (Mumby et al. 2004) when mangroves are not available. Living in these 'interconnected habitat mosaics' (*in sensu* Sheaves 2005) therefore places limits on the level of direct dependence of fish and crustaceans on mangroves. Simply, a given species may be crucially

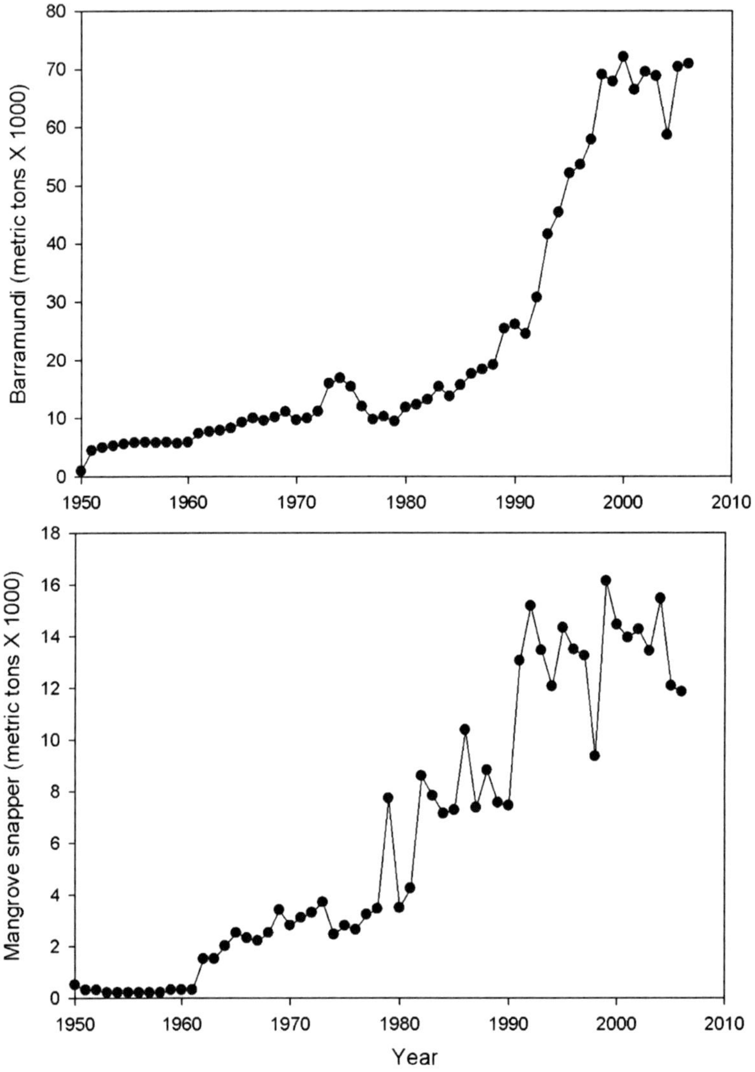

Fig. 4.8 Annual landings of wild barramudi (*Lates calcarifer*) and mangrove red snapper (*Lutjanus argentimaculatus*) caught in Southeast Asian waters. (Data from 2007 FAO Fishstat Plus software http://www.fao.org/fishery/topic/16073)

dependent on a small patch of mangrove habitat for only a brief period, but this crucial and tenous link is unlikely to surface in catch data and the subsequent statistical analyses. The brief residency of a species will make it difficult to detect why it depends on mangroves.

Table 4.6 lists the known statistical correlations of mangrove area with either shrimp or fish abundance, or both. Lee's analysis (2004) using a principal components- regression

Table 4.6 Correlations of mangroves with shrimp and fish

Region	Relationship	Correlation (r^2) (n)	Reference
Tropics worldwide	Shrimp-total intertidal vegetation area	0.54 (27)	Turner (1977)
New World tropics	Shrimp-total intertidal vegetation area	0.64 (14)	Turner (1977)
Indonesia	Shrimp-mangrove area	0.89 (NA)	Martosubroto and Naamin (1977)
Gulf of Carpentaria, Australia	Shrimp-mangrove length	0.58 (6)	Staples et al. (1985)
Gulf of Mexico	Fish-mangrove area	0.48 (10)	Yãnez-Arancibia et al. (1985)
Tropical worldwide	Shrimp-mangrove area	0.53 (NA)	Pauley and Ingles (1986)
Peninsular Malaysia	Shrimp-mangrove area	0.89 (10)	Sasekumar and Chong (1987)
Philippines	Shrimp-mangrove area	0.61 (18) 0.66 (18)[a]	Paw and Chua (1991)
Philippines	Fish-mangrove area	0.34 (15)– 0.66 (12)[a]	Paw and Chua (1991)
Philippines	Shrimp and fish-mangrove area	0.40 (34)– 0.45 (39)	Paw and Chua (1991)
Vietnam	Shrimp and fish-mangrove area	0.95 (NA)	de Graaf and Xuan (1998)
Vietnam	Shrimp-mangrove area	0.88 (5)	de Graaf and Xuan (1998)
Tropics worldwide	Shrimp-multiple[b]	0.38 (37)	Lee (2004)
NSW, Australia	Fish-total mangrove, salmarsh, seagrass	0.32–0.75 (49)	Saintilan (2004)
Malaysia	Shrimp-mangrove area	0.37–0.70 (36)	Loneragan et al. (2005)
Queensland, Australia	Shrimp-mangrove area	0.37–0.70 (36)	Manson et al. (2005b)

NA = not available; [a] = one shrimp or fish species only; [b] = coastline length, rainfall, temperature, latitude, tidal amplitude, absolute and relative mangrove area.

approach suggests that the extent of intertidal area and organic matter availability as represented by tidal amplitude, rather than just total mangrove area, function as the main drivers for shrimp production. He found that latitude, tidal amplitude, absolute and relative mangrove area, coastline length, rainfall, and temperature account for 78% of the overall variation in shrimp catch data. The other studies (Table 4.6) all point to a significant relationship of mangrove area with either shrimp or fish catch, or both, but don't explain why this is so; a correlative approach does not take into account the underlying mechanisms or the cause-and-effect relationships of the link between mangrove and fisheries yields. All that can be stated at this time is that estuarine residents, both temporary and permanent, are dependent on mangroves for myraid reasons (known only to the shrimp and fish!).

Chapter 5
The Forest Floor

5.1 Introduction

Pelagic processes such as plankton productivity and the transport of dissolved and particulate detritus are functionally important, but it is on and within the forest floor that many of the most essential energetic processes and trophic relations within mangroves take place. Many epibenthic, root epibiont, and infaunal organisms in mangrove forests harvest a wide range of foods—from DOM to bacteria and fungi to macroalgae to amorphorus detritus to wood. It is this catholicity that makes it so difficult of categorize benthic biota trophically; the problem of separating biota from fine soil particles is what makes it so difficult to categorize them energetically. In this chapter, we will examine life in and on the forest floor, with the main focus on the most energetically significant group, the soil microbiota.

5.2 Soil Composition and Physicochemical Attributes

Mangrove soil is commonly described as smelly, sticky, anoxic mud—a nuisance to be endured when entering the tidal forest. In truth, the forest floor can vary from dead hard corals and boulders to quartz or carbonate sand, to very fine silt and clay. There is therefore no 'typical' soil profile, as they can be either acidic or alkaline (pH range: 5.8–8.5), anoxic or suboxic (redox range: −200 to +300 mv). Soil texture and associated physicochemical properties depend on constituent trees and roots, parent rock, geomorphology, tides, and rainfall (Alongi 2005a; Ferreira et al. 2007b). The more quiescent the location the more likely that silt and clay will deposit, imparting a fine texture to the soil. In large deltaic, mature forests, significant quantities of peat and fine, fibrous roots make up the soil, whereas in forests fringing river banks, soils are sandy. The smaller the grain size, the more DOM and POM the soil contains; concentrations of POC and total nitrogen can range from 0.5–15% and 0.2–0.5% by soil dry weight, respectively. Soils constitute the largest reservoir of organic and inorganic elements within mangrove forests (see Section 2.2.3). If mangrove soils can be said to be unique, two somewhat typical features that are important to plant–soil–microbe relations make them so: (1) concentrations of dissolved inorganic nutrients, consisting mostly of ammonium, phosphate, and nitrate are in the low μM range, and (2) soluble and condensed tannins leached from roots and litter, produced by the trees for chemical defense against herbivores impregnate the soil. Tannins can be a large fraction of the interstitial DOC pool (Alongi 2005a).

The chemistry of the interstitial water is greatly influenced not only by tree physiology and microbial decomposition, but also by tidal range and frequency of inundation, salinity, temperature, redox, pH, and other physical factors (McKee 1993). Replenishment by tides and dilution by rain water are significant controlling factors in many forests as more frequently air-exposed soils are subject to percolation and dilution of interstitial water by monsoonal rains, as tidal advection alternatively pumps large quantities of seawater and air into and out of the soil (Alongi 2005a).

pH and redox of mangrove soils are controlled by a constellation of factors: the composition and concentration of the inorganic and organic components, tidal elevation, water content, microbial activity, grain size, and extent of anthropogenic input, if any. In a number of mangrove forests, soil pH is <7.0, attributable to oxidation of iron sulfides promoted by the release of oxygen from roots, especially in the surface layers (Gleason et al. 2003; Marchand et al. 2003) as well as by production and release of organic acids from mangrove roots and metabolic by-products of microbial decomposition of organic matter. Mangrove soils are suboxic to anoxic below a thin veneer of oxidized soil (5–10 mm) on the surface of the forest floor and in burrow linings, cracks and fissures, but they are rarely sulfidic because the buildup of free sulfides is often prevented by (1) oxygen translocated to the roots, (2) active mixing of surface and subsurface soils via bioturbation, and (3) pumping and flushing of tidal waters (Thongtham and Kristensen 2003). Some infauna, such as crabs and mudskippers, entrap air in their burrows (Ishimatsu et al. 1998; Thongtham and Kristensen 2003) and even ant colonies that nest in mangrove muds concentrate metabolic gases in pockets below-ground (Nielsen et al. 2006).

Soil properties specific to different mangrove species have been suggested by a number of studies showing species-specific differences in redox status, dissolved oxygen, particulate nutrient and free sulfide concentrations, and degree of pyritization (McKee et al. 1988; McKee 1993, Lacerda et al. 1995; Alongi et al. 1998, 1999, 2000a, b; Alongi 2001). Evidence points to higher concentrations of organic matter and more anoxic conditions in *Rhizophora* soils than in *Avicennia* soils. In Brazilian forests, *Rhizophora* soils have higher concentrations of organic carbon and nitrogen than *Avicennia* soils. A similar situation exists in forests of Western Australia, where redox potential is more reducing in *Rhizophora* soils (Alongi et al. 2000a). A number of studies (Table 5.1) show apparent species-specific differences in soil redox status, although with significant seasonal and depth variations, most differences are not statistically significant. The only valid comparisons are those made between species in a given locale. For example, in comparisons between *R. stylosa* and *A. marina* forests at three different locations in Western Australia, redox potential was significantly less in the *Rhizophora* forest at two locations, but at the third site, the opposite was true, with more anoxic conditions in the *Avicennia* forest (Alongi et al. 2000a). Also, redox differences among the three *Rhizophora* forests were as different as they were among the three *Avicennia* forests. Species-specificity of soil characteristics can therefore be overstated.

Forest age may be an important factor in soil maturity, especially accumulation of organic matter via the buildup of dead roots. A plot of soil total organic carbon content with age of various *Rhizophora apiculata* forests in Vietnam, Malaysia, and Indonesia points to a significant positive relationship between net accumulation of organic matter and stand age (Fig. 5.1). Along the coast of French Guiana, in close proximity to the Amazon, Marchand et al. (2003) examined the composition of soil organic matter among various successional stages of *Avicennia germinans* forests. They found a trend toward increasing soil organic content and C/N ratio as follows: 1-m high pioneering forest → 4-m high recolonized forest → 10-m high young forest → 20-m high mature forest. These findings are similar to often observed

Table 5.1 Two examples of apparent species-specific differences in redox potential (Eh, mv) in forests of Western Australia and Micronesia. Values are mean ±SE of replicate cores

Species	Eh (mv)	Location	Reference
R. stylosa	−103 ± 36	Western Australia	Alongi et al. (2000a)
A. marina	195 ± 93		
A. corniculatum	10 ± 27		
B. gymnorrhiza	75 ± 27	Micronesia	Gleason et al. (2003)
S. alba	−5 ± 27		
R. apiculata	21 ± 28		

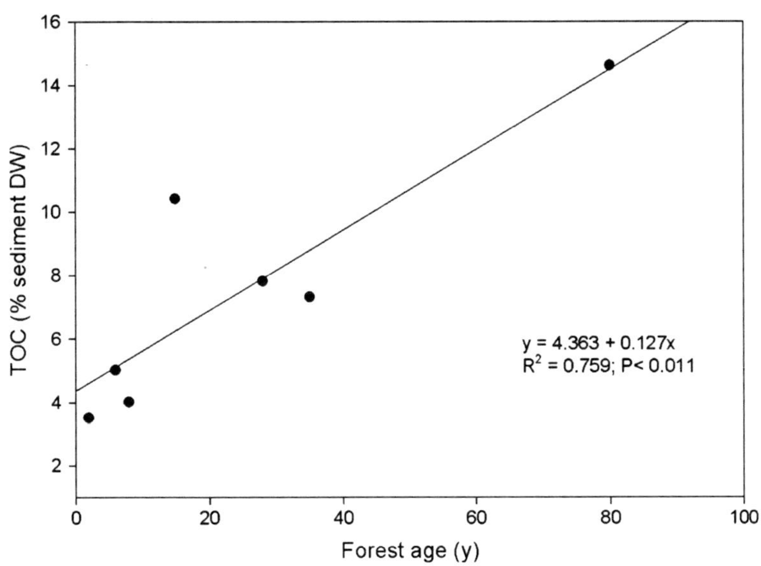

Fig. 5.1 The relationship between soil total organic carbon content and age of various *Rhizophora apiculata* forests in Southeast Asia (Data from Alongi et al. 1998, 2000b, 2001, 2008)

patterns of organic matter accumulation in terrestrial forest soils (Perry et al. 2008). Presumably, the same ecological factors operate in both mangroves and upland forests. As it is not known at this stage if forests of other mangrove species exhibit similar increases in soil organic content with increasing stand age, we cannot state that this is a universal pattern for mangroves.

The origin and composition of organic matter in mangrove soil does change with increasing stand age (e.g., Marchand et al. 2003). A number of studies have subjected mangrove soils to sophisticated organic chemical analyses to determine their origin and chemical composition. Using both chemical and optical analysis, Marchand et al. (2003) found an increase in the proportion of organic matter attributable to development of the radial cable root system of *A. germinans*. There was an increase in ligno-cellulosic debris with stand age, but there were also qualitative differences in the content of carbohydrates; algal- and microbial-derived compounds were more common in the youngest forests, corresponding to the presence of algal mats. In deeper deposits of the oldest forests, the chemical composition suggests increased proportions of oxidized allochthonous debris derived from a legacy of detrital discharge from the Amazon (Marchand et al. 2005).

The chemical analysis of soil organic matter more often reflects selective degradation by microbes rather than the source(s) of the organic material. Even chemical markers resistant to degradation during early diagenesis, such as pentacyclic trierpenols derived from mangrove leaf waxes, can be difficult to interpret because signatures can be mixed and diluted or there can be a substantial change in isotopic composition during diagenesis (Koch et al. 2005). The organic matter of mangrove soils is a rich mixture of mangrove litter, dead and live roots, wood and mangrove peat, with allochthonous contributions from seagrasses, reef algae, up-river terrestrial soils and plants, animal wastes, microbial biomass, and marine POM (Muzuka and Shunula 2006). A recent analysis of stable isotope data (Kristensen et al. 2008) for mangrove soils suggests that although there is a distinctive signal present of mangrove litter, there is in some cases a large input of organic matter derived from other autotrophs and terrestrially-derived debris. Mangrove soils are therefore a mixture of inorganic and organic ingredients derived from an array of land- and ocean-based sources, with proportional contributions dependent on geomorphological history and location.

5.3 Life on the Forest Floor

A virtual menagerie of organisms is encountered walking, running, crawling, sliding, swimming, and hiding among leaves, twigs, roots, bark, timber, mounds, tubes, and burrows on the forest floor (Alongi and Sasekumar 1992; Kathiresan and Bingham 2001). A variety of herbivores, omnivores, and carnivores also occupy various heights within the canopy (Section 2.6). All of these organisms form complex communities tightly interlinked to the trees (Cannicci et al. 2008), either directly or indirectly, exhibiting often quite complex patterns of zonation across the intertidal seascape, and with height above the forest floor and mean sea-level. Many of these patterns, while reflecting responses to physical and chemical factors, are also the result of intense competition and predation (Schrijvers and Vincx 1997).

Gastropods and crustaceans are the major invertebrates living on the soil surface; their patterns of distribution and abundance in relation to frequency of tidal inundation, competition, and predation are well-described (Alongi and Sasekumar 1992; Nagelkerken et al. 2008). Less understood are their rates of respiration, ingestion, egestion, and secondary production. For instance, there is an abundant literature on prop root epifauna such as sponges, barnacles, isopods, bryozoans, tunicates, polychaetes, hydroids, mollusks, and arthropods, regarding their species composition and to a lesser extent, their rates of recruitment (Section 2.5.4), but no secondary production estimates. This is also true for most wood-boring fauna, including the dominant teredinid mollusks. Nearly all fauna and flora living on or above the forest floor are fed upon by fishes at high tide but the significance of these activities from an energy and material flow perspective is almost wholly unknown. Most information on energetics focuses on the benthic organisms that process the litter carpeting the forest floor, as well as decomposition dynamics of the litter.

5.3.1 The Role of Crabs in Consumption of Seeds and Litter

A paradigm shift occurred in the mid-1980s regarding the principal flow of energy and materials within mangrove food chains. Based on work in south Florida, most leaf litter was earlier thought to be transported into tidal creeks and waterways (the 'outwelling' hypothesis, Section 6.2) where the material was decomposed by microbes along the route of mangrove litter → saprophytes → detritivores and omnivores → lower carnivores → higher carnivores (Heald 1969; Odum and Heald 1975). Subsequent studies conducted in the Indo-Pacific, the Caribbean, and South America showed, however, that a variable but large proportion of litter reaching the forest floor is consumed or hidden below-ground mostly by crabs (Robertson et al. 1992a), thereby reducing the amount of material available for export and serving as a nutrient retention mechanism.

In the Indo-Pacific, the studies of Smith (Smith 1992) and Robertson (Robertson et al. 1992a) and their colleagues established that crabs of the sub-family Sesarminae directly consume both leaf litter and propagules; herbivory on the latter has a major impact in determining the structure and species composition of many forests. Smith (1987a, b) hypothesized that the amount of seed predation is inversely related to the dominance of a given tree species. Smith and his colleagues (Smith et al. 1989b) then performed a series of experiments in Australia, Malaysia, Florida, and Panama to test the dominance-predation hypothesis on propagules of several species of *Avicennia, Bruguiera*, and *Rhizophora*. They found that *Avicennia* fit the model, with seed predation inversely correlating with the relative abundance of tree species in the canopy. Cumulative amounts of predation were different among four species of *Avicennia*, but the amounts were high, ranging from a low of 46%

for *A. officianalis* to a high of 72% for *A. germinans*. The results were mixed for *Rhizophora* and *Bruguiera*, although the cumulative amounts of seed predation were less for these genera (range: 0–24% for *Rhizophora*, 5–40% for *Bruguiera*). Higher rates of predation on *Avicennia* can be attributed to their greater nutritive value and lower tannin content.

The lack-of-fit to the dominance-predation hypothesis (in most instances) for *Rhizophora* and *Bruguiera* is attributable to differences in composition of the predator guilds. In manipulative experiments conducted in Panama, Sousa and Mitchell (1999) similarly found that the propagules of *A. germinans* are more heavily preyed upon than those of *Rhizophora mangle* and *Laguncularia racemosa*, but concluded that while predation patterns on seeds of *A. germinans* are consistent with the dominance-predation hypothesis, the rates of predation are insufficient to exclude the species from the low intertidal zone. The other two mangrove species did not fit the model, indicating that crab predation on propagules is not a universal explanation for species zonation across the intertidal.

Further experiments indicate either that seed predation greatly influences mangrove species distribution (e.g., McGuiness 1997; Lindquist and Carroll 2004) or does not conform to the dominance-predation model (Clarke and Kerrigan 2002). Clarke and Kerrigan (2002), for instance, measured predation rates ranging from 22–100% of seed abundance with predation rates as follows: *Aegiceras corniculatum* > *Avicennia marina* > *Bruguiera parviflora* > *Aegialitis annulata* > *B. exaristata* > *Ceriops australis* > *C. decandra* = *B.gymnorrhiza*. Again, these species differences can be attributed to differences in tannin and nutritional content.

Most grapsid crabs consume litter, but a few species eat fresh leaves on the tree. These fresh leaf consumers can climb as high as the tops of trees. These species include *Sesarma leptosoma* in the Indo-Pacific, *Aratus pisonii* in the Neotropics, and *Sesarma elegans* in West Africa (Erickson et al. 2003). In Brazilian forests, *Aratus pisonii* consumes green leaves equivalent to 2–5% of total leaf area, and prefers leaves of *Laguncularia racemosa* which have higher nitrogen and phosphorus content that leaves of *Rhizophora mangle* and *Avicennia schaueriana* (Faraco and da Cunha Lana 2004). But while these species consume significant quantities of fresh leaves from individual trees, the bulk of leaf consumption in most tidal forests is carried out on leaves carpeting the forest floor.

The proportions of litter consumed or buried underground by different species of crabs in different mangrove forests (Table 5.2) indicates that while consumption rates are usually quite high, not all are. For example, within a *Sonnertia alba* forest in Kenya, there was no significant leaf consumption (Table 5.2). Furthermore, in some temperate Australian forests and in some stands of the Caribbean, crabs seem to avoid eating mangrove leaves and seeds (Saintilan et al. 2000; Guest et al. 2004, 2006).

Variation in the level of consumption relies on the relative frequency of tidal inundation and nutritional quality of the litter, in addition to the species of mangrove and predator. Competition between crabs and other benthic consumers also plays a role in the extent of leaf litter consumed. In mangrove stands in Gazi Bay, Kenya, the snail *Terebralia palustris* and the grapsid crab *Neosarmatium smithi*

Table 5.2 Rates of litter consumption (% total litter stock) by various crab species and other benthic animals

Consumer	Mangrove	Consumption rate (%)	Location	Reference
Perisesarma onychophorum, P. eumolpe	*R. apiculata, R. stylosa*	9–30	Malaysia	Leh and Sasekumar (1985)
Perisesarma messa	*R. apiculata, R. stylosa, R. lamarckii*	28	Australia	Robertson (1986)
Perisesarma bidens, Parasesarma affinis	*Kandelia candel*	57	Hong Kong	Lee (1989)
Perisesarma messa, Neosarmatium smithi	*Ceriops tagal*	71	Australia	Robertson and Daniel (1989a)
Perisesarma messa, Neosarmatium fourmanoiri	*Bruguiera exaristata*	79	Australia	Robertson and Daniel (1989a)
Neosarmatium fourmanoiri, Parasesarma moluccensis	*Avicennia marina*	33	Australia	Robertson and Daniel (1989a)
Neoepisesarma spp., *Perisesarma* spp.	*R. apiculata*	>100	Thailand	Poovachiranon and Tantichodok (1991)
Neosarmatium meinerti	*A. marina*	44	South Africa	Emmerson and McGwynne (1992)
Parasesarma guttatum	*Rhizophora mucronata*	19	Kenya	Slim et al. (1997)
Perisesarma onychophorum, P. eumolpe	*R. apiculata, Bruguiera gymnorrhiza, B.parviflora. Avicennia officinalis*	79	Malaysia	Ashton (2002)
Neosarmatium meinerti	*A. marina*	>100	Kenya	Olafsson et al. (2002)
Neosarmatium meinerti	*Sonneratia alba*	0	Kenya	Olafsson et al. (2002)
Ucides cordatus (ocypodid)	*Rhizophora mangle*	61–81	Brazil	Schories et al. (2003), Nordhaus et al. (2006)
Terebralia palustris (gastrod)	*C. tagal*	11	Kenya	Slim et al. (1997)

compete directly for *R. mucronata* leaves lying on the soil surface. The snails systematically forage for leaves *en masse*, while the crabs drag leaves into their burrows to reduce or avoid competition for food with the snails; crabs pull leaves away from snails but only if they are large enough and only if few snails are on the leaf (Fratini et al. 2000). Snails can be significant leaf consumers (Table 5.2) when crabs are not present.

The voracity of many grapsid and ocypodid crabs for mangrove litter seems paradoxical. Why consume litter that is of low nutritional value? Some initial studies suggested that perhaps the crabs paste litter fragments to their burrow walls to allow microbial colonization and sufficient decomposition to make the material more palatable and nutritious (Lee 1998). This idea was indirectly supported by observations that most crab species preferred and grew better on senescent rather than on fresh leaves, or preferred leaves of low tannin content. Subsequent work with stable isotopes and detailed analyses of excavated burrows reveals that the stable carbon isotope signature of some sesarmids resembles the C signal of mangrove soil rather than that of leaves, with no difference in nutrient content between senescent leaves and leaf fragments lining burrow walls (Skov and Hartnoll 2002). This evidence implies that leaves do not provide adequate amounts of nitrogen required for crab growth and reproduction. Skov and Hartnoll (2002) have suggested that crabs supplement their meager diet of litter by ingesting soil detritus. Another possibility is that they consume carrion and other invertebrates, such as meiofauna (Thongtham and Kristensen 2005; Lee 2008).

Shredding, ingesting, and assimilating litter plays an important role in facilitating decomposition of the material. The pulmonate gastropod *Melampus coffeus* consumes mangrove litter on the forest floor, but escapes up tree trucks to avoid the incoming tide at Boca Ceiga Bay in Florida (Proffitt and Devlin 2005). Consuming nearly 41% of leaf fall, populations of *Melampus coffeus* greatly increase litter breakdown with litter bag experiments showing a 90% weight loss over one month for *A. germinans* litter and over 7 weeks for *R. mangle* leaves, compared to litter bags without gastropods present. Similar results for litterbag experiments were recorded by Middleton and McKee (2001) in Belizean island forests, indicating a tripling of litter breakdown in the presence of crabs and amphipods, but little decay of more refractory litter, such as twigs and roots. Consumption of leaf litter facilitates microbial colonization by producing material, including feces, with a greatly increased surface area to volume ratio, thus acting as a positive feedback loop in the flow of energy and materials within mangrove food webs.

A large number of studies have examined the reproductive histories of many species of mangrove crustaceans, including crabs, but estimates of secondary production of mangrove crabs and other benthic organisms are few. Macintosh (1977, 1984) estimated production of several sesarmid and fiddler crab species in Malaysian mangrove forests and calculated annual production rates of 3.1–16.2 g m^{-2} $year^{-1}$ for three species of *Uca* and 9.1–9.7 g m^{-2} $year^{-1}$ for two species of sesarmids. The secondary production of *Uca lactea annulipes* in an Iranian mangrove estuary (Mokhtari et al. 2008) was estimated at 1–3.8 g DW m^{-2} $year^{-1}$ with males accounting for 80% of the productivity. In a more detailed study in a mangrove forest and adjacent creek banks of the Caeté estuary, Koch and Wolff (2002) calculated rates of secondary production for various crab species of 6.5–147.9 g AFDW m^{-2} $year^{-1}$ (Table 5.3). The detritivous crab guild dominates in terms of secondary production and respiration and the high net production efficiency (mean = 19%) suggests that

Table 5.3 Mean rates of secondary production, respiration, assimilation, and net production efficiency (NPE = production/assimilation ratio) in several species of crabs and other epibenthos occupying mangrove forest and small and large creek banks in the Caeté estuary, northern Brazil (Data from Koch and Wolff 2002)

Habitat and species	Production (g AFDW m^{-2} $year^{-1}$)	Respiration (g AFDW m^{-2} $year^{-1}$)	Assimilation (g AFDW m^{-2} $year^{-1}$)	NPE (%)
Forest				
Uca rapax	6.0	33.6	39.9	15
Uca vocator	10.5	31.8	42.3	25
Uca cumulanta	1.6	6.5	8.1	20
Ucides cordatus	2.6	62.6	65.2	4
Small creek				
U. cumulanta	7.2	20.2	27.4	26
Uca maracoani	4.7	14.0	18.7	25
Pachygrapsus gracilis	3.7	26.1	29.8	12
Eurytium limnosum	1.3	7.7	9.0	14
Thais coronata	0.01	1.0	1.01	1.0
Large creek				
U. maracoani	53.4	147.9	201.3	27
T. coronata	0.15	3.2	3.35	5

the foods consumed are highly nutritious. Koch and Wolff (2002) also postulated that these high efficiencies and the fact that the crabs assimilated organic matter equivalent to 15% of forest net primary production reflect a positive feedback loop between the crabs, soil bacteria, and the trees. It thus appears that there is a network of positive feedback loops in mangrove forests to enhance nutrient recycling and to help conserve nutrients.

5.3.2 *Patterns of Microbial Decomposition of Litter*

Crabs and other benthic detritivores initially process nearly 50% of the total litterfall produced in many forests, but what happens to the rest of the material? A variable proportion is swept away by the tides (see Chapter 6), but any litter remaining in the forest is then further decomposed by various microbial flora; litter shredded but unassimilated by the crabs is returned to the soil as waste and is also consumed by microbes. In many cases, litter that has not been immediately processed by macro-consumers is initially degraded by microbes.

Detritus not processed by macroconsumers decomposes in three stages: (1) leaching of soluble compounds, (2) saprophytic decay, and (3) fragmentation. A large number of studies measured the microbial decay of mangrove leaf detritus (Robertson et al. 1992a; Kristensen et al. 2008) and several common features emerge:

- Absolute decay parameters are site- and species-specific.
- Leaves decompose faster in subtidal areas than in the intertidal zone.
- Leaf decay is more effective when leaves retain their moisture.
- Leaves of species such as *Avicennia* and *Kandelia* with lower tannin content and higher initial nitrogen concentrations decompose faster than *Rhizophora* and *Bruguiera* leaves.
- Leaf decomposition of the same species occurs at similar rates in both the tropics and subtropics, but more slowly in the dry tropics where leaves are subject to intense aridity and high salinity.

The initial stage of leaching involves the loss, on average, of 20–40% of the organic carbon in the leaf when submerged for 10–14 days. The first components to be lost are the non-ligncellulose carbohydrates such as sugars, tannins, and other phenolic compounds. This leachate is quite labile and is readily degraded aerobically and incorporated into microbial biomass with conversion efficiencies up to 90% (Kristensen et al. 2008). The remaining particulate detritus then undergoes the second phase of decomposition with the colonization by aerobic and anaerobic prokaryotes, and oomycotes (zoosporic members of the Protoctista). Ascomycetes (true fungi) appear to play a relatively minor role in mangrove litter colonization (Newell 1996); it is the oomycotous protoctists, especially *Halophytophthora vesiculara*, that are especially well-adapted to capture cellulosic compounds via pervasion and digestion, to the extent that these polysaccharide components are decomposed about twice as fast as the lignins. Mangrove detritus thus becomes relatively enriched in lignin-derived carbon over time. Lignin has a half-life of >150 years, so it is degraded only very slowly, especially under anoxic conditions when incorporated in the soil.

The chemical changes that occur in decomposing mangrove detritus is a function not only of enzymatic degradation and loss of leachable compounds, but of compositional changes in the microbes that have colonized the material. Accumulation of microbial biomass over years of decomposition results in a relative enrichment of N and P relative to C, but also an increase in hydrolyzable amino acids, total amino acids, bacterial biomarkers, lipids, and monosaturated and branched chain fatty acids typical of bacteria (Mfilinge et al. 2003; Tremblay and Benner 2006). Tremblay and Benner (2006) found that after 4 years of decomposition, leaves of *Avicennia germinans* are composed of 60–75% N and 20–40% C derived not from the original plant tissues, but from heterotrophic bacteria. Relative phosphorus content in decomposing leaves of *Rhizophora apiculata* increases 174–220% of the initial amount in the litter after several months, probably in concert with binding of P to humic compounds and iron (Nielsen and Andersen 2003). These chemical studies help to explain the early findings of a general decrease in the C:N and C:P ratios of decomposing mangrove detritus over time.

The decay constants of yellow leaves deposited onto mangrove soil range from 0.001–0.1 day^{-1} with decomposition usually following the single exponential pattern, $M_t = M_0 e^{-kt}$, where M_t is the mass remaining after time t, M_0 is the initial mass, and k is the decay constant (Kristensen et al. 2008). This progression of decay is controlled by several factors such as mangrove species, initial C:N:P ratio, tidal inundation frequency, and abundance of detritivores. Wetting is a particularly crucial factor to faciliate leaching of soluble compounds and accessibility by microbes. The rate of decay is, of course, also greatly augmented by shredding by leaf-eaters, such as crabs and amphipods (Kristensen and Pilgaard 2001; Bosire et al. 2005). A significant inverse relationship exists between the decay constant and the C:N ratio of mangrove leaf litter (Fig. 5.2). This inverse correlation implies that the rate of microbial decomposition of mangrove leaf litter is strongly dependent on the availability of nitrogen relative to carbon.

Leaves from genera such as *Avicennia* and *Kandelia* decompose much faster than litter originating from species with more structural lignocellulose, such as *Rhizophora* (Fig. 5.2). In addition to species differences in degradability and elemental composition, there are also differences between locations which can be attributed to regional differences in nutrient availability or forest age. Overall, however, decomposition rates and nitrogen immobilization of mangrove litter is controlled by the initial chemical composition of the residues. Perhaps, as in terrestrial forests (Manzoli et al. 2008), decomposers in mangrove forests lower their carbon-use efficiency to exploit litter of low initial N content.

The decomposition of leaf litter has important trophic implications beyond the forest floor. For example, a peptic inducer originating from bacteria decomposing mangrove leaves cues settlement and metamorphosis of planula larvae of the tropical jellyfish *Cassiopea xamachana* (Fleck and Fitt 1999). Planulae prefer to settle on

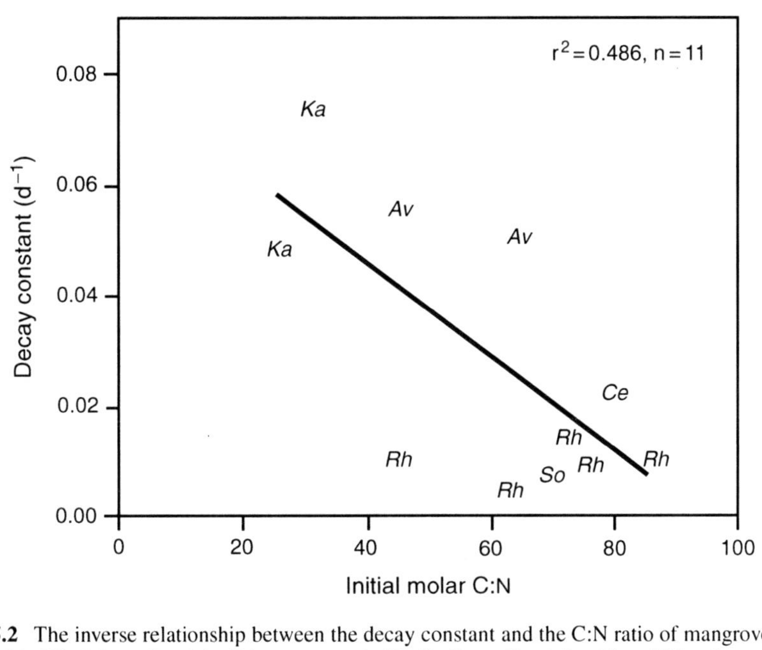

Fig. 5.2 The inverse relationship between the decay constant and the C:N ratio of mangrove leaf litter (Modified from fig. 4 in Kristensen et al. 2008). Ka = *Kandelia*, Rh = *Rhizophora*, So = *Sonneratia*, Ce = *Ceriops*, Av = *Avicennia*.

the shady side of leaves and reject other substrates for settlement. Whether or not gregarious settlement of other marine invertebrates is induced by chemical cues released from leaves or other components in mangrove forests is unknown, but this work again illustrates how processes in the forest and in tidal waters are linked.

5.3.3 Crabs as Ecosystem Engineers

Burrowing grapsid and ocypodid crabs usually dominate the benthic infauna of mangroves and have considerable impact on ecosystem functioning to the extent that they have recently been labeled 'ecosystem engineers' (Kristensen 2008). 'Ecosystem engineers' are organisms that directly or indirectly exert a modifying or controlling influence on the availability of resources to other species by causing changes in the physical state of the environment. In Section 3.2.2 we saw how crab burrows alter the flow and transport of water and suspended materials from the forests to adjacent waterways, and in the previous Section (5.3.1), how they consume a significant fraction of litter that falls to the floor of many forests. Both roles have an decisive impact on nutrient retention within the ecosystem, thus minimizing the outwelling of such material.

Five processes are crucial in considering crabs as ecosystem engineers (Cannicci et al. 2008; Kristensen 2008):

- Changes in soil texture and porosity induced by their burrowing and maintenance activities.
- Redistribution of materials (fluids, gases) by reworking within the soil or active pumping of material in and out of the soil via burrows.
- Handling and ingesting potential food resources and by doing so modifying and changing soil redox.
- Burrows mediating transport of substances via diffusive gas transport, passive fluid transport as well as collapsing burrows.
- Change in the reactive states of substances via removal of toxic metabolites (e.g., H_2S) and introduction of oxygen deep into the soil.

Sesarmid and fiddler crabs construct burrows in mangrove soils to retreat from intolerable environmental conditions (e.g., high tide, desiccation, high temperatures) and to use as a refuge from competition and predation. Fiddler crabs prefer open spaces, such the banks of mangrove creeks, where sunlight is sufficient to foster surface growth of benthic microalgae. Sesarmids on the other hand prefer to burrow into soil under the canopy where litter is most abundant and where the shaded soil and associated roots provide protection against high temperatures and predators. Neither crab group actively ventilates its burrows or secretes mucus to directly stimulate microbial activity. Instead, they increase microbial activity and the transport of solutes and gases by their sporatic crawling movements and by constructing burrows of such design as to increase the interface between soil and the water/air interface.

Aside from transporting litter down their burrows, crabs alter the biogeochemistry of soil lining the burrow wall by facilitating oxygen penetration, which increases the volume of oxidized Fe^{3+} soil and neutralizes anaerobically-derived metabolites. There is an increased supply of reactive Fe^{3+} in soils of the burrow walls tipping the delicate redox balance in favor of iron reduction over sulfate reduction (Kristensen 2008). Smaller amounts of toxic sulfides are therefore produced and what H_2S is produced is quickly converted to relatively inert pyrite or may be re-oxidized back to sulfate.

Field experiments in which sesarmid crabs were excluded reveal some profound effects not only on sediment chemistry but also on forest productivity (Smith et al. 1991). In *Rhizophora*-dominated forests in north Queensland, Australia, Smith et al. (1991) found that exclusion of sesarmid crabs results in increased levels of sulfide and ammonium in the porewater compared with control plots in which crabs are left undisturbed. Perhaps more importantly, stipule fall is significantly less in the exclusion plots as is mangrove reproductive output. This experiment underscores the crucial role of crabs in influencing the productivity of mangrove forests.

Sesarmid crabs also contribute to the rates and pathways of microbial carbon oxidation in mangrove soils by producing fecal pellets that sustain high microbial activity and acting as a food source for detritus-feeders, such as amphipods (Lee 1997). The feeding activities of fiddler crabs are different, being restricted to the upper few mm of soil, but these small crabs exert a clear impact on microbial activities just as much as sesarmids do via their foraging behavior (Meziane et al. 2002). In a microcosm experiment, Kristensen and Alongi (2006) examined the impact of the presence/absence of the fiddler crab *Uca vocans* on microbial fluxes with and without saplings of *Avicennia marina*. The saplings grew more leaves and pneumatophores in the presence of the crabs. When crabs were absent from microcosms, dense microalgal mats developed leading to stimulation of benthic primary production and respiration (Fig. 5.3). Lowest respiration rates were

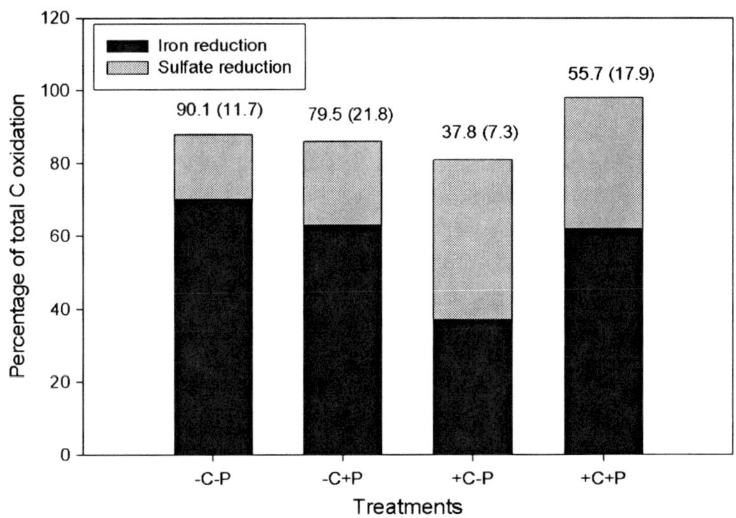

Fig. 5.3 Comparison of the impact of the presence (+)/absence (–) of fiddler crabs and mangrove saplings on rates and pathways of microbial biochemistry (Data from Kristensen and Alongi 2006)

measured in microcosms with crabs but without saplings; this result was due to crab grazing significantly reducing the standing stock of surface microalgae in turn reducing the available supply of labile carbon for microbial decomposition. The percentage of microbial decomposition that occurred via the sulfate reduction pathway was greatest in this treatment as leaching of DOC from mangrove roots stimulated the activity of sulfate reducers. Fe^{3+} was the most important electron acceptor for microbial carbon mineralization in ungrazed soil (63–70%) whereas SO_4^{3-} was more important in grazed sediment (36–44%), especially in the presence of *Avicennia* (Fig. 5.3).

5.3.4 Trophic Dynamics of Other Macrobenthos

The diets of other benthic macrofauna have been elucidated by a variety of methods, such as gut analysis and feeding experiments, but recent studies using stable isotopes and fatty acid biomarkers have unravelled the dietary mysteries of many invertebrates (Table 5.4). The food of many epifauna and infauna has been classified as detritus when, in reality, most gut contents are unidentifiable or consist of organic matter from an unknown source. Fatty acid and stable isotope analyses (Table 5.4) show to a remarkable degree the lack of a distinctive mangrove signal; a more common characteristic of these invertebrates is a signature suggesting foods derived from other autotrophs, bacteria, and organic matter plus some plankton. In a series of detailed studies of benthic invertebrate diets using stable isotope results, Bouillon and his colleagues (Bouillon et al. 2002a, b, 2004) found that in Indian mangrove estuaries, mangrove-derived organic matter is not the principal food source, but that phytodetritus from the waterways is the main food for most species; only a limited number of species assimilate mangrove-derived organic matter. In the pre-monsoon season, pelagic and benthic microalgae are the preferred foods and rely almost entirely on microalgal carbon as a food source in the post-monsoon season.

The mudskipper, although not an invertebrate, is one of the faunal icons of mangrove forests, occupying a specialized niche in the intertidal zone, and living an amphibious lifestyle in intimate contact with other benthic organisms (Clayton 1993). Mudskippers are carnivorous, feeding mostly at low tide on prey picked from the soil surface, but until recently the contribution of various food items to their diet has remained illusory. A detailed analysis of the mudskipper *Periophthalmus argentilineatus* inhabiting the mangrove shoreline in Tanzania found that dietary switching occurs from juvenile to adult stages (Kruitwagen et al. 2007). Fish up to 60 mm in total body length feed mostly on small copepods and amphipods. Larger individuals growing as long as 70–110 mm shift their prey selection to polychaetes and thereafter to a diet of mangrove crabs. Stable isotope signatures confirm these gut results and show that the prey items are derived from local mangroves.

Benthic carnivores and detritivores are abundant in most mangrove forests, but a new paradigm is emerging, in that many of the most conspicuous soil-dwelling fauna feed on a wide variety of foods, especially those of algal origin. Mangrove

Table 5.4 Diets of benthic macroinvertebrate species found in a variety of mangrove forests and waterways worldwide as deduced by fatty acid biomarkers and stable isotopes. FA = fatty acid; SI = stable isotope; B = both methods

Species	Diet	Method	Location	Reference
Austrovenus stutchburyi (clam)	Dinoflagellates, diatoms, plant detritus	B	Matapouri, New Zealand	Alfaro et al. (2006)
Paphies australis (clam)	Dinoflagellates, diatoms, plant detritus	B	Matapouri, New Zealand	Alfaro et al. (2006)
Crassostrea gigas (oyster)	Dinoflagellates, diatoms, plant detritus	B	Matapouri, New Zealand	Alfaro et al. (2006)
Turbo smaragdus (gastropod)	Brown algae, bacteria, diatoms, zooplankton	B	Matapouri, New Zealand	Alfaro et al. (2006)
Nerita atramentosa (gastropod)	Brown algae, bacteria, zooplankton	B	Matapouri, New Zealand	Alfaro et al. (2006)
Lepsiella scobina (gastropod)	Diatoms, zooplankton, dinoflagellates	B	Matapouri, New Zealand	Alfaro et al. (2006)
Cominella glandiformis (gastropod)	Plant detritus, diatoms, brown algae, zooplankton	B	Matapouri, New Zealand	Alfaro et al. (2006)
Palaemon affinis (shrimp)	Diatoms, plant detritus, dinoflagellates, brown algae	B	Matapouri, New Zealand	Alfaro et al. (2006)
Neanthes glandicincta (polychaete)	Benthic microalgae	SI	Taiwan	Hsieh et al. (2002)
Laonome albicingillum (polychaete)	Pelagic POM	SI	Taiwan	Hsieh et al. (2002)
Batillaria zonalis (gastropod)	Green macroalgae, bacteria, diatoms	FA	Okinawa, Japan	Meziane and Tsuchiya (2000)
Terebralia sulcata (gastropod)	Bacteria, green macroalgae, mangrove litter	FA	Okinawa, Japan	Meziane and Tsuchiya (2000)
Cerithideopsilla cingulata (gastropod)	Bacteria, diatoms, mangrove litter, green macroalgae	FA	Okinawa, Japan	Meziane and Tsuchiya (2000)
Geloina coaxans (mud clam)	Mangrove detritus, bacteria	FA	Okinawa, Japan	Bachok et al. (2003)
Onchidina australia (gastropod)	Benthic microalgae	SI	Queensland, Australia	Guest and Connolly (2004)

detritus is still a key food resource for some organisms, but it probably plays an equally important role in nutrient recycling and retention to help sustain high rates of net canopy production in many forests. Until recently, the trophic role of micro- and macroalgae has been greatly underestimated.

5.3.5 Wood Decomposition

Mangroves, like all trees, eventually die, fall, and come to rest on the forest floor where they are quickly colonized by fungal colonies and other organisms that either live within the forest or are imported with the tides (Cragg 1993; Allen et al. 2000; Maria and Sridhar 2004). But it is the teredinid mollusks that are the main agents of wood decomposition, possessing symbiotic cellulolytic and nitrogen-fixing bacteria that assist in breaking down wood. The onset of wood boring by the mollusks does not begin until after significant leaching of dissolved matter has occurred. A laboratory screening of natural resistance in tropical hardwoods has found that several mangrove species produce leachates that induce mortality in marine borers, with the leachate of all species being toxic; *B. gymnorrhiza* being especially so compared with *Heritiera littoralis, R. stylosa*, and *Xylocarpus granatum* (Borges et al. 2008). After this initial leaching stage, decomposition by marine borers speeds up. Robertson and Daniel (1989b) found that after nearly 16 years lying on the forest floor decaying, trunks of *Rhizophora* were riddled with a network of tubes produced by teredinids. Some workers (Chai 1982; Ong et al. 1984; Hauff et al. 2006) noted more rapid breakdown of logs lying within some Malaysian and Micronesian forests, whereas 9–10 years after the passage of Hurricane Andrew, 66% of coarse woody debris was decomposed on the floor of a south Florida mangrove forest (Krauss et al. 2005).

Detritus from wood decomposition can be almost as important as leaf litter consumption by crabs in some forests (Fig. 5.4). In mature and young *Rhizophora* forests in northern Australia, Robertson and Daniel (1989b) estimate that the flux of fallen wood and crab consumption rates of litter are nearly equivalent in the mature forest, whereas leaf consumption is the main detrital pathway in the young forest. They found that trunks decompose rapidly (k = 0.108 year^{-1}) compared with wood in terrestrial forests, but branch wood decomposes even more rapidly (k = 0.302 year^{-1}). After nearly 16 yr, the trunks still retained 20% of their original carbon, although the C:N ratio dropped from 1,400 to 190 during the first year of decomposition. Branches retained 50% of their original carbon in a 2.5 year experiment with the C:N ratio declining from 125 to ≈90 (Fig. 5.4).

The rate of wood decomposition varies in relation to mangrove species, position of the wood (i.e., lying down or standing), and intertidal location. Using small wood disks cut from live stems of *Avicennia germinans, Laguncularia racemosa*, and *Rhizophora mangle* and placing them up above tidal waters, on and beneath the soil surface, Romero et al. (2005) studied their decomposition patterns in a south Florida estuary for 28 months. Decay in the air followed a simple exponental

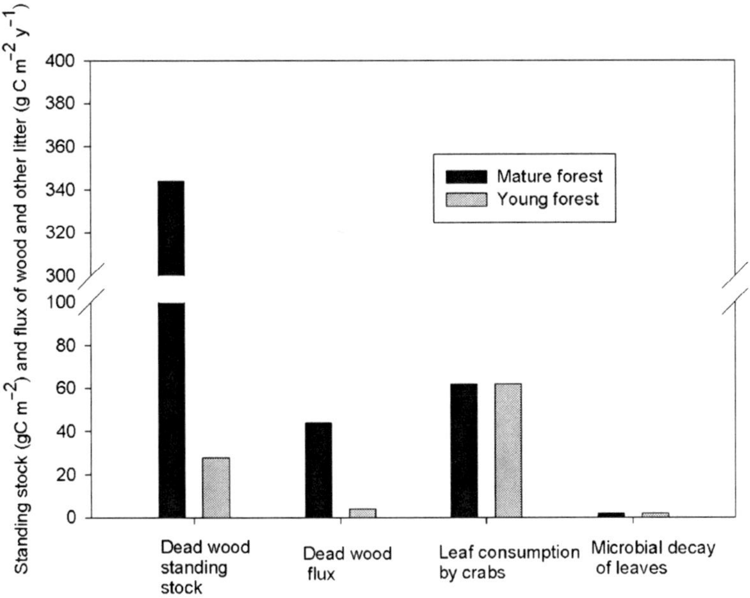

Fig. 5.4 Estimates of the mass of fallen wood and the fluxes of wood detritus, leaf consumption by crabs and rates of microbial decay of leaves in mature and young *Rhizophora* forests in northern Australia (Data from Robertson and Daniel 1989b)

model with an effect of location and species. However, disks buried in the soil and on the surface decomposed following a two-component model for labile and more refractory components. Labile components decomposed at rates of 0.37–23.71% month^{-1} with *A. germinans* decomposing most quickly and *L. racemosa* decomposing most slowly. Surface disks decomposed faster than buried ones, and both sets of disks decayed more rapidly with the disks in the canopy. The three species showed similar overall rates of decay. N content increased in surface and buried disks but 17–68% of P was lost via leaching during the first 2 months of the experiment. Phosphorus content remained nearly constant for the remainder of the study.

5.3.6 Root Decomposition

As with root production (Section 2.2.1), few workers have studied root decomposition in mangrove forests (Table 5.5). This is despite the fact that roots comprise a significant, if highly variable, fraction of soil mass in many forests and appear to contribute significantly to forest production. The first published studies were conducted in *Avicennia marina* forests in New Zealand (Albright 1976; Van der Valk and Attiwill 1984). The latest (and only other) studies were conducted by McKee

Table 5.5 Rate of below-ground fine and coarse root decomposition (% root mass lost day^{-1}) from New Zealand and Belizean forests

Species	Root diameter (mm)	Placement in soil	Decay (% day^{-1})	Location	Reference
Avicennia marina	1	Buried	0.19	New Zealand	Albright (1976)
Avicennia marina	1	Surface	0.34	New Zealand	Albright (1976)
Avicennia marina	1	Buried	0.06	New Zealand	Van der Valk and Attiwill (1984)
Avicennia marina	10–20	Buried	0.22	New Zealand	Van der Valk and Attiwill (1984)
Avicennia germinans	<2, >10	Buried, low/mid intertidal zone	0.108	Twin Cays, Belize	Middleton and McKee (2001)
Avicennia germinans	<2, >10	Buried, high intertidal zone	0.104	Twin Cays, Belize	Middleton and McKee (2001)
Rhizophora mangle	<2, >10	Buried, low/mid intertidal zone	0.108	Twin Cays, Belize	Middleton and McKee (2001)
Rhizophora mangle	<2, >10	Buried, high intertidal zone	0.092	Twin Cays, Belize	Middleton and McKee (2001)
Rhizophora mangle	≤2.5	Buried, low intertidal zone	0.17	Twin Cays, Belize	McKee et al. (2007)
Rhizophora mangle	>2.5	Buried, low intertidal zone	0.08	Twin Cays, Belize	McKee et al. (2007)
Rhizophora mangle	≤2.5	Buried, mid intertidal zone	0.17	Twin Cays, Belize	McKee et al. (2007)
Rhizophora mangle	>2.5	Buried, mid intertidal zone	0.08	Twin Cays, Belize	McKee et al. (2007)
Rhizophora mangle	≤2.5	Buried, high intertidal zone	0.167	Twin Cays, Belize	McKee et al. (2007)
Rhizophora mangle	>2.5	Buried, high intertidal zone	0.065	Twin Cays, Belize	McKee et al. (2007)

and her colleagues in mangroves of Belize (Middleton and McKee 2001; McKee et al. 2007). In Twin Cays, Belize, rates of root decomposition for *A. germinans* range from 0.104–0.108% day^{-1} with slightly slower decomposition rates in the high intertidal zone. For roots of *R. mangle*, there were no significant differences in decay rates with intertidal position, but decomposition of large coarse roots were slower (range: 0.065–0.08) than rates of fine root breakdown (range: 0.167–0.17). In the *R. mangle* study, McKee et al. (2007) found that addition of N and P fertilizers had no effect of the rate of fine and coarse root decomposition. In the Florida Everglades, similarly slow rates of decomposition were measured, but Poret et al. (2007) concluded that soil conditions (nutrient content, degree of tidal inundation) had a greater effect on root decomposition that root quality.

These studies imply that rates of below-ground root decomposition are slow compared with breakdown of other tree components, such as leaves. The slow decay of mangrove roots explains the formation of peat in many mangrove forests, as rates of input must greatly exceed rates of output and decay of organic material in order for peat to accumulate (Middleton and McKee 2001; McKee 2001). Peat formation is thus another mechanism to store and retain essential elements, including refractory carbon and nitrogen, within mangrove ecosystems.

5.4 Microbial Processes in Forest Soils

Bacteria in soils are the essential players in the flow of energy and materials in ecosystems; only the trees rival these prokaryotic assemblages in mangrove energetics. Bacteria, along with other microflora such as flagellates, ciliates, and amoebae, are the crux of the microbial hub in mangrove soils, metabolizing organic matter and serving as food for other benthic organisms, such as meiofauna. Various methods have been used to estimate the abundance, biomass, and metabolic activities of soil microbes. Many of these methods are no longer in vogue, but sufficient information exists from biogeochemical studies to offer a clearer picture of the energetic role of microbes in soil nutrient transformation processes and in forming links to the trees.

It is now recognized that the often high rates of mangrove primary productivity depend not only on unique and highly evolved physiological mechanisms, but also on highly evolved and efficient interrelationships among soil nutrient pools, microbes, and trees (Section 2.5). Close links are especially crucial in the tropics because available nutrient pools (e.g., nitrate) are small and turn over rapidly, and microbial growth is rapid in consistently warm temperatures. However, because of the inherent difficulties in separating microbiota from soil particles, there is more inferential data than direct evidence of the complexity and nature of these interrelationships.

Bacteria, fungi, and protists alter the microenvironment around mangrove roots via their metabolic activities, transforming and releasing nutrients, and modifying soil chemistry (Holguin et al. 2001). The relationship between microbes and trees is paradoxically both competitive and mutualistic, as both trees and microbes share, and share the need for, limiting nutrients. Nutrient-use efficiencies of mangroves are equal to or higher than those of other tropical trees (Section 2.5.4) implying that bacterial transformations of nutrients and subsequent tree growth are rapid.

Mangrove–microbe relations have been most closely observed within the rhizosphere where highly specialized groups of Archaea, bacteria, protists, and fungi coexist within the root matrix (Sengupta and Chaudhuri 1991; Ravikumar et al. 2004; Kothamasi et al. 2006). In Indian mangroves, Kothamasi et al. (2006) found arbuscular mycorrhizal fungi in the aerenchymatous cortex of several mangrove species, suggesting that the plants may be providing the fungi with oxygen; phosphate-solubilizing bacteria were also abundant suggesting that these bacteria mobilize insoluble phosphates for the plants. Mangrove trees alter the soil environment, and this affects the growth and survival of individual functional types of aerobic and anaerobic bacteria. It has been known for decades that translocating oxygen to the roots serves as a means of oxidizing potentially toxic metabolites, such as sulfides. It has been shown only recently that these activities shift the competitive balance for substrates from favoring sulfate-reducers to favoring iron- and manganese-reducing bacteria, thus increasing availability of soluble Fe and Mn required for plant growth. Highly evolved and energetically efficient plant–soil–microbe relations are a major factor in explaining why mangroves are highly productive in harsh tropical environments.

5.4.1 *Rates and Pathways of Bacterial Decomposition of Soil Organic Matter*

The bacterial decomposition of organic matter in marine sediments and saline soils follows a sequence based on availability of electron acceptors which is related to vertical changes in redox chemistry and population abundance of the various metabolic types of bacteria. Where oxygen is present, aerobic respiration occurs, giving way to the suboxic and anoxic pathways of denitrification, manganese reduction, iron reduction, sulfate reduction, and methanogensis (Fig. 5.5). The vertical sequence is really not so simple, as sediments and soils have microzones in which some of these metabolic processes co-exist. What all of these processes have in common is the production of CO_2, and this can be measured at the soil surface in enclosed chambers to give an estimate of total bacterial decomposition of carbon. Acetate produced via fermentation of particulate organic carbon is the main utilizable form of carbon fueling these metabolic processes (Kristensen et al. 1994).

Rates of both dissolved and gaseous flux of oxygen and carbon dioxide across the soil surface are thought to be measures of total carbon decomposition. Organic matter oxidation results in the production of CO_2, and O_2 consumption in the dark is an equally good measure of total soil respiration. However, only aerobic and microaerophilic prokaryotes and other heterotrophs, such as most protists and all invertebrates, use oxygen. However, most anaerobic pathways result in the production of reduced metabolites (e.g., H_2S), most of which gets oxidized when diffused to the soil surface. So, nearly all anaerobic metabolic output should be indirectly accounted for by measuring consumption of oxygen in the dark.

A compilation of available data on oxygen consumption and carbon dioxide production (Table 5.6) shows that flux rates tend to be higher when soils are exposed to the atmosphere than when they are immersed by tides. This reflects the fact that molecular diffusion is faster for gases than fluids and that during air exposure when

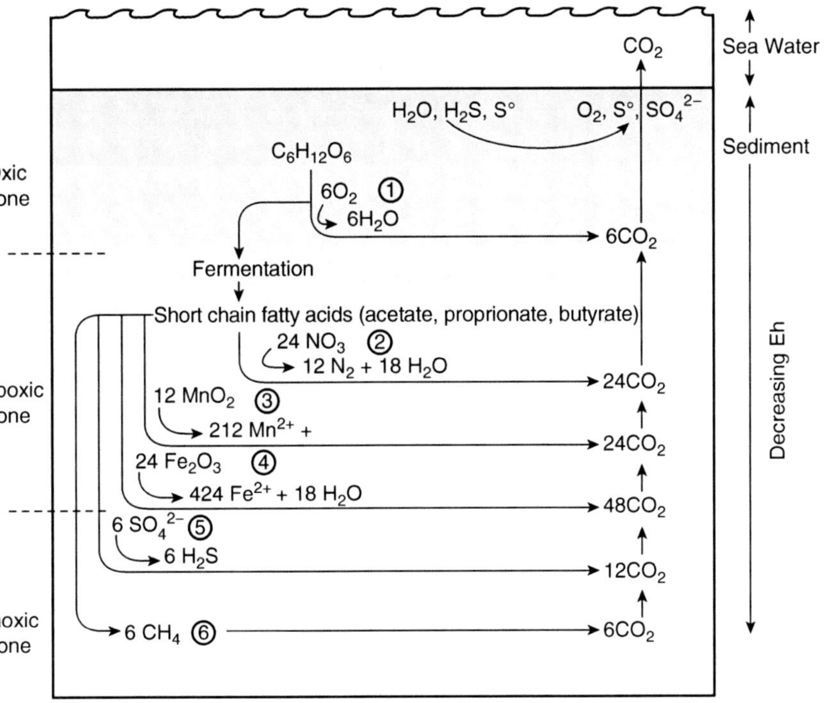

Fig. 5.5 Schematic representation of the sequence of bacterial metabolic processes of organic carbon in mangrove soils

Table 5.6 Mean rates of oxygen consumption and CO_2 production (mmol m^{-2} day^{-1}) and the respiratory quotient ($RQ = CO_2/O_2$) measured in tidally-inundated and air-exposed soils from various mangrove forests worldwide (Data from Kristensen et al. 1988, 1992, 2000; Nedwell et al. 1994; Middelburg et al. 1996; Alongi et al. 1998, 1999, 2000a, b, 2001, 2004a, 2005b, c, 2008; Holmer et al. 1999. Earlier references are in Alongi 1989)

	Mean	Standard error	Number of observations
O_2			
Inundated	35.93	4.86	55
Exposed	64.57	11.08	58
All	50.63	2.20	113
CO_2			
Inundated	49.32	6.29	62
Exposed	68.96	8.27	75
All	60.07	5.42	137
RQ (CO_2/O_2)			
Inundated	1.63	0.13	52
Exposed	1.32	0.27	53
All	1.47	0.15	105

fissures, cracks, and burrow openings become replenished with air, the surface area increases for aerobic respiration and chemical oxidation. This may also partly reflect warmer temperatures when soils are exposed to tropical air.

Higher rates of carbon dioxide than oxygen flux results in overall respiratory quotients slightly >1 (Table 5.6), suggesting that the flux of these gases and solutes from the soil surface reflects decomposition of close to model Redfield ratio organic matter ($C_{106}H_{260}O_{106}N_{16}P_{1}$) or marine microalgae or phytoplankton ($C_{106}H_{177}O_{37}N_{17}S_{0.4}P_{1}$), both of which yield respiration coefficients for complete oxidation of 1.3 and 1.45, respectively (Middelburg et al. 2005). However, measurement of benthic respiration across the soil surface may represent mainly decomposition of organic matter in surface deposits rather than representing an overall measure of total organic matter decomposition in the forest floor. As discussed later in this chapter, when individual pathways of carbon oxidation are measured and summed the product is often greater than the rates of O_2 and CO_2 flux measured across the soil surface (Alongi et al. 2001, 2004a, 2005b). This means that a significant proportion of bacterial decomposition is not being accounted for by respiratory measurements at the soil surface; lateral advection and tidal drainage may result in the loss of respired carbon from deep, subsurface processes within the forest floor and transported to adjacent waterways. The forest floor is therefore reminiscent of a giant 'sponge' which readily gains and loses interstital water and atmospheric gases when pumped by the tide.

So, what do measurements of surface soil respiration relate to? A number of individual studies (see the references in Table 5.6) found significant correlations of respiration with soil temperature, redox potential, organic carbon and nitrogen content, grain size, and mangrove net primary productivity, but there appears to be no single overarching factor that regulates soil respiration. Using the data in Table 5.6 and the accompanying soil data in these references, only one significant, but weak, correlation was found between CO_2 production and soil N:P ratio (Fig. 5.6). Given the large matrix, it is just as likely that the relationship is spurious rather than indicating nitrogen limitation in relation to phosphorus availability.

An attempt to relate carbon dioxide release with forest and edaphic characteristics across a broad degree of latitude (27° N to 37° S) found that mangrove soil respiration and Q_{10} (2.6) was similar to those measured in terrestrial forests, correlating with leaf area index and litterfall (Lovelock 2008). The correlations, though significant, were weak as no one factor controls soil respiration in mangrove forests.

Differences in respiration rates between forests of different species have been observed. For example, several workers have measured greater respiration rates in *Rhizophora* than in *Avicennia* soils, but as noted earlier, it is not known why this is so. In species comparisons between forests in Western Australia, Alongi et al. (2000a) found that greater respiration rates in *Rhizophora* forests coincided with smaller grain size and higher organic content. However, the *Rhizophora* forests were located lower in the intertidal zone, so the species differences may just as likely be due to inherent differences in tidal inundation and subsequent accumulation rates of soil organic matter than to inherent species-specific factors.

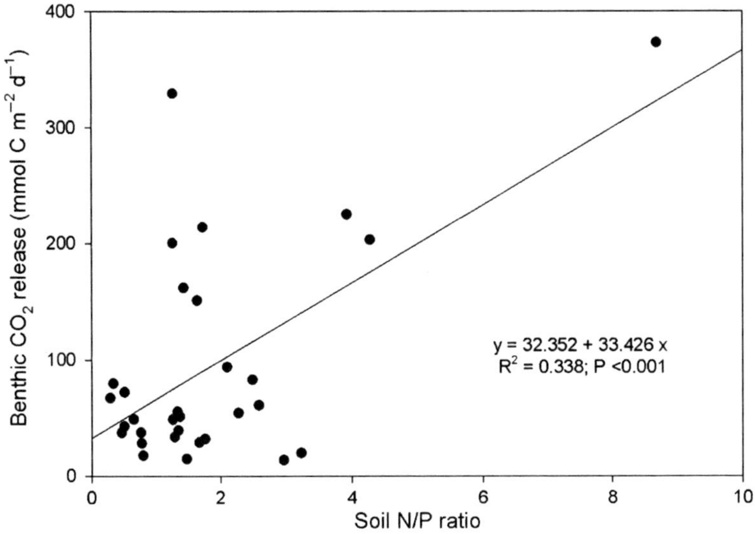

Fig. 5.6 The relationship between benthic CO_2 release and soil N/P ratios in mangrove forests worldwide. References in Table 5.6

The most realistic statement that can be made regarding soil respiration is that different factors dominate in different forests, and may even differ within the same forest over time. There is enormous heterogeneity in soils, including patchiness of benthic fauna, which can have an impact on rates of metabolism (Kristensen and Alongi 2006; Kristensen 2007). The release of gases and other metabolites from tree roots undoubtedly plays a role in regulating rates of soil metabolism but it is unlikely that rates of surface soil metabolism are closely linked to forest production (Lovelock 2008) as earlier studies indicated that the percentage of carbon lost via soil respiration in mangrove forests is usually low (Alongi 2005a).

5.4.2 *Sulfate Reduction*

Aerobic respiration and anaerobic sulfate reduction are usually the major decomposition pathways in mangrove soils (Alongi 2005a; Kristensen 2007). As oxygen is depleted below the upper few mm of surface soil (excluding soils lining burrows, cracks and fissures), anaerobic metabolism dominates, with decomposition of organic matter mediated by fermenting and sulfate-reducing bacteria. Most mangrove soils therefore are characterized by high concentrations of reduced inorganic sulfur, especially pyrite (FeS_2) and elemental sulfur (S°), but negligible levels of iron monosulfides (FeS) and free sulfide (H_2S).

Short-term incubations with radiolabelled $^{35}SO_4$ for determination of rates of sulfate reduction usually result in the recovery of a large fraction of the reduced radiolabel in a distilled form of pyrite, and to a much lesser extent, elemental sulfur. Significant recovery of radiolabel in acid-volatile FeS and HS^- occurs in rapidly accreting soils and/or when accompanied by rapid accumulation of labile organic matter, such as organic farm wastes and sewage (Alongi et al. 2005b, c). The oxidizing activity of roots, burrowing activities of crabs and other infauna, and low pH favors rapid formation of pyrite either through direct precipitation of Fe^{2+} with polysulfides or via FeS oxidation with elemental sulfur and polysulfides (Holmer et al. 1994). The rate of pyrite formation may be limited by the availability of reactive iron, most of which is bound in pyrite. The storage of pyrite in mangrove deposits varies with soil type and depth, which often vary with tidal height. Mangrove peat deposits contain a considerable amount of pyrite as organically bound sulfur.

Oxidation of reduced sulfur may be an important recycling pathway for inorganic sulfur in mangrove soils; oxidation of FeS_2 is mediated either directly or indirectly by iron oxides (Holmer et al. 1994; Kristensen and Alongi 2006). In addition to simple molecular diffusion, oxygen may be transported to deeper soils by release from roots, by mixing by crabs and other benthos, and by advective transport in surface layers by wave and tidal action. Pyrite oxidation can also be caused by human disturbance of the soil profile, resulting in serious acidification when forests are cleared for aquaculture and other development. Soil disturbance can also disrupt the natural sequence of bacterial metabolism, resulting in a diminution of anaerobic metabolism (Alongi and de Carvalho 2008).

Using the references in Alongi (2005a) and Kristensen et al. (2008), an average rate of sulfate reduction from 96 measurements was calculated: 36.2 mmol S m^{-2} day^{-1} with a standard error of 6.1, a median value of 12.9, and a range of 0.2–319.0 mmol S m^{-2} day^{-1}. On average, the rates of sulfate reduction in mangrove soils appear to be lower than rates measured in many salt marshes (Canfield et al. 2005). However, most estimates of sulfate reduction in mangroves are underestimates as only the upper 5–10 cm of soil was sampled in most cases. In some mangroves, sulfate reduction is detectable to soil depth exceeding 1 m (e.g., Alongi et al. 2001). Comparing these values with soil nutrient content, grain size, and temperature, no significant correlations was found across forests. Like total soil respiration, no one factor regulates sulfate reduction in mangrove soils.

Spatial and temporal patterns in sulfate reduction have been identified in a number of mangrove forests. For example, in *Kandelia candel* forests in China, rates of sulfate reduction declined significantly from the low to the high intertidal, mirroring rates of mass sediment accumulation and soil nutrient content (Alongi et al. 2005b). Other studies found no clear or consistent pattern across the intertidal, although most found some seasonality that related to rainfall (or the lack thereof) rather than temperature. For instance, in Thailand, Alongi et al. (2001) measured higher rates of sulfate reduction during the wet southwest monsoon than during the dry season when soils were particularly dry at low tide.

An analysis of sulfate reduction rates with the age of various *Rhizophora apiculata* forests throughout Southeast Asia (Fig. 5.7) shows some decline in sulfate reduction in forests older than about 35 years. This relationship suggests that other metabolic

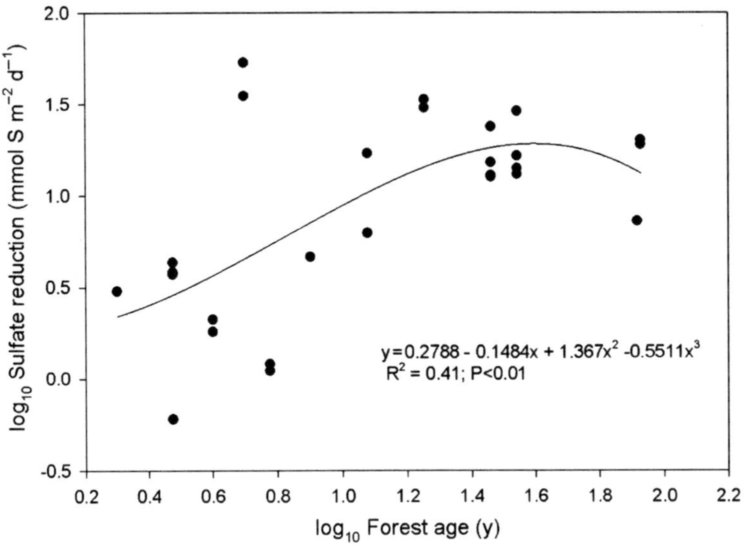

Fig. 5.7 Log_{10}–log_{10} relationship between rates of bacterial sulfate reduction in mangrove soils and age of *Rhizophora apiculata* forests in Southeast Asia (Data from Alongi et al. 1998, 2000b, 2001, 2004a, 2005b, c, 2008

pathways, such as iron and manganese reduction, may exceed rates of sulfate reduction, but it may also suggest that rates of anaerobic metabolism decline in concert with the leveling off or the beginning of the decline in *Rhizophora* forest production (Fig. 2.16). Most of the variations in sulfate reduction at a given forest age represent seasonal patterns and inherent site-specific variations; these forests vary in location from marine to brackish water, and from low to mid intertidal. A significant positive relationship between sulfate reduction and net primary productivity among *Rhizophora stylosa* and *Avicennia marina* forests in Western Australia was found, but the age of these forests is not known (Alongi et al. 2000a).

The most likely direct connection between forest age (or net primary production) and rate of sulfate reduction is the exudation of dissolved solutes from roots and their uptake and assimilation by sulfate-reducers. In forests of southern Thailand, rates of sulfate reduction correlated significantly with biomass of both live and dead fine roots (Alongi et al. 2001), and these relationships were attributed to uptake and utilization of exudates from live roots and dead root matter (e.g., lignins; Dittmar and Lara 2001a). The clearest evidence that mangroves directly impact sulfate reducing activity in soils was found by Kristensen and Alongi (2006) in mesocosm experiments with and without crabs and saplings of *Avicennia marina*. They found greater rates of sulfate reduction in mesocosms with plants and found that leaching of DOC from roots stimulated sulfate reduction and an increase in bacterial abundance. There is thus a direct functional link between sulfate-reducing bacteria and mangrove roots.

5.4.3 Iron and Manganese Reduction

The activites of sulfate reducers are closely intertwined with the presence of iron- and manganese-reducing bacteria. The mesocosm experiments of Kristensen and Alongi (2006) with *A. marina* indicate complementary effects of both crabs and saplings on the metabolic processes in mangrove soils. While rates of sulfate reduction were greater in the presence of saplings, the exudation of DOC from roots had a cascading effect on iron dynamics near the soil surface. All Fe^{3+} was efficiently reduced and converted into the solid Fe^{2+} pool and all sulfide generated by sulfate reducers was rapidly precipitated as iron sulfides. Absence of Fe^{2+} in the upper few cm of the soil must have been caused by rapid reoxidation and precipitation as amorphous iron oxides. Fe^{3+} was the most important electron acceptor in mesocosms without crabs, accounting for 63–70% of total microbial carbon decomposition. It is highly likely that this proportionally large contribution of iron reducers was due to the presence of saplings or to the presence of benthic microalgae.

The rates of iron reduction in these experiments (equivalent to 20.6–63.4 mmol C m^{-2} day^{-1}) are comparable to rates measured in natural mangrove soils (Kristensen et al. 2000). In mangroves and intertidal mudflats on Phuket Island, Thailand, Kristensen et al. (2000) measured rates of iron reduction ranging from 5.2–36.1 mmol C m^{-2} day^{-1}. The highest rates were found in a *Rhizophora mucronata* forest where Fe^{3+} reduction accounted for 70–80% of total microbial carbon decomposition; reduction rates were lower (as were the proportional contributions) in the non-vegetated tidal flat sediments. This pattern can be attributed to the stimulatory effects of roots and bioturbating infauna, with grain size, iron and nutrient content in the soil considered important co-factors. Similar results were obtained in Brazilian mangroves (Ferreira et al. 2007a).

In Indian mangroves, measurable rates of net iron and manganese release were detected in close association with tree roots (Alongi et al. 2005c). Alongi et al. (2000a, b) measured net release of dissolved Fe and Mn from incubated soils in other forests and found that metal reduction is usually not an important process. However, the methodology seriously underestimated the actual rates of metal reduction, especially manganese reduction. In mangrove soils of the Bahamas, rates of manganese oxidation range from 3–119 pmol mg soil DW^{-1} h^{-1}, with highest rates in the lowest salinity waters (Spratt and Hodson 1994) suggesting that manganese reduction may be an important process in less saline soils where low sulfate levels limit the activity of sulfate-reducing bacteria.

In a recent review, Kristensen (2007) noted a positive correlation between the pools of reactive Fe^{3+} and the proportion of iron respiration contributing to total microbial carbon decomposition in mangrove soils. He postulated that when the concentration of reactive Fe^{3+} exceeds ≈35 μmol cm^{-3}, more than 80% of the anaerobic carbon oxidation is mediated by iron reducers. Further studies are needed to clarify the circumstances in which metal reduction pathways take on increased importance in microbial metabolism in mangrove soils. This knowledge is also important to further understand the function of iron plaques that build up on mangrove roots. Such plaques may serve to render toxic metals harmless, but this idea is only conjecture (Machado et al. 2005).

5.4.4 Methane Release

Due to their low energy yield, methanogenic bacteria are inferior competitors with other anaerobic bacteria for electron donors such as hydrogen and acetate. Processes such as sulfate reduction can keep concentrations of hydrogen and acetate at levels too low for methanogens. Thus, methanogensis is restricted to soils and sediments where electron acceptors such as nitrate, sulfate, and metal oxides have been exhausted. Methanogenesis therefore accounts for only a small fraction (1–10%) of total microbial carbon decomposition in mangrove soils. Methanogenic activity has been detected in some, but not all, mangrove soils (Alongi 2005a; Kristensen 2007). Interest in measuring methane fluxes from mangrove soils has increased over the past decade owing to the fact that methane is a greenhouse gas with a global warming potential 7–62 times greater than carbon dioxide.

Kristensen (2007) recently summarized the available data and concluded that methane production is usually low and highly variable, and cannot be detected in some forests (Giani et al. 1996; Alongi et al. 2000a, 2001, 2004a). Emission rates, when measurable, normally range from 0.1–5.1 mmol CH_4 m^{-2} day^{-1}. In highly polluted mangroves, rates can reach nearly 60 mmol CH_4 m^{-2} day^{-1} (Verma et al. 1999) and in subtropical mangroves, maximum rates can approach 30 mmol CH_4 m^{-2} day^{-1} in summer (Allen et al. 2007). Although methanogenesis is usually detected in deeper soil strata, some studies have detected methane in surface soils, especially in estuarine forests where low salinities mean lower concentrations of sulfate, limiting rates of sulfate reduction (Lyimo et al. 2002; Lekphet et al. 2005). Methane release has been detected in soils where sulfate reduction has also been measured; this coincidence probably reflects microenvironments that allow coexistence or the presence of methylated amines that are used only by methanogenic bacteria (Canfield et al. 2005).

The rate of methane emission is most closely related to the degree of organic enrichment, with higher rates in organically-enriched soils (Giani et al. 1996; Purvaja and Ramesh 2000, 2001; Strangmann et al. 2008) and, to a lesser extent, by seasonal changes in soil temperature and soil moisture (Lu et al. 1998; Ye et al. 1999; Allen et al. 2007). Domestic wastes induce severe oxygen stress and supply labile organic carbon in sufficient quantities to stimulate methanogensis (Sotomayor et al. 1994). Logging can also indirectly stimulate methane release, probably because of lower redox potentials due to less oxygen being pumped into the soil by live roots (Giani et al. 1996). Experiments by Strangmann et al. (2008) indicate that elevated methane concentrations and fluxes in polluted soils reduce growth of mangrove seedlings, suggesting one mechanism why seedlings may not grow well in soils polluted by organic enrichment from aquaculture and sewage. Although considered a minor process, methanogens may have a larger role to play in regulating initial colonization or re-introduction of pioneering mangroves.

Methanogenesis occurs not just in soils, but also in and on parts of trees. Prop roots of *Rhizophora stylosa* emit CH_4 as discovered by Kreuzwieser et al. (2003) in Australian forests, where emission rates range from 3.9–5.0 μmol CH_4 m^{-2} root surface day^{-1}. The pneumatophores of *Avicennia marina* are also colonized by methanogens (Purvaja et al. 2004) as emission rates correlate positively with the density of pneumatophores. Methanogens live and grow in the aerenchymatic tissues; CH_4 concentrations decline from the base of the pneumatophores to the emergent tip (Purvaja et al. 2004).

Mangrove waterways can also be significant sources of methane (Barnes et al. 2006; Ramesh et al. 2007). Obviously, the roots of other mangrove species need to be examined for methanogenic activity. In both pristine and polluted mangrove ecosystems in India, Barnes et al. (2006) and Ramesh et al. (2007) found that tidal waters are supersaturated in CH_4 with respect to atmospheric concentrations, with emission rates ranging from 3.3–10.4 mmol CH_4 m^{-2} day^{-1} in pristine waters to a maximum of 5,216 mmol CH_4 m^{-2} day^{-1} in highly polluted mangroves. Mangroves may therefore be a greater source of methane to the atmosphere than previously believed.

5.4.5 Nitrogen Processes and Links to Trees

Nitrogen is usually limiting to estuarine and marine ecosystems, and this is also true for many mangrove forests. The vast bulk of nitrogen is taken up via fine roots in the form of nitrate and/or ammonium, and it is for this reason that an understanding of soil nitrogen transformations is very important. It is therefore surprising that the complete (or nearly so) soil N cycle has been studied in only three mangrove forest ecosystems: on Phuket Island (Kristensen et al. 1995, 1998, 2000) and in Sawi Bay in southern Thailand (Alongi et al. 2002) and in Missionary Bay on Hinchinbrook Island in northern Australia (Alongi et al. 1992; Alongi 2005a). Individual processes such as denitrification have been measured numerous times in a number of forest locations, but only when the pieces of the nitrogen puzzle have been put together can the role of soil nitrogen in the ecosystem be clarified.

Nitrogen budgets for low- and mid-intertidal forests of *Rhizophora apiculata* on Phuket Island (Fig. 5.8) show that most of the organic nitrogen in the soil is broken down to NH_4^+ by ammonifying bacteria with proportionally little loss to the atmosphere via denitrification. Nitrogen fixation is not a major input, but algae are important, as most solute flux across the soil–water interface is taken up within surface mats. Only about 5% of total nitrogen input is buried in the soil. Most NH_4^+ is presumably taken up by tree roots as uptake of dissolved N from the water column only accounts for 9–10% of N required for mangrove net primary production.

In the mangrove soils of four forests of Sawi Bay (Fig. 5.9), most nitrogen similarly flows via the ammonium pool, as N burial and denitrification equate to only 4–12% and 3–23% of total N input. The turnover time of the soil ammonium pool is 7–41 hours so, by difference, 70–90% of the porewater NH_4^+ must be taken up

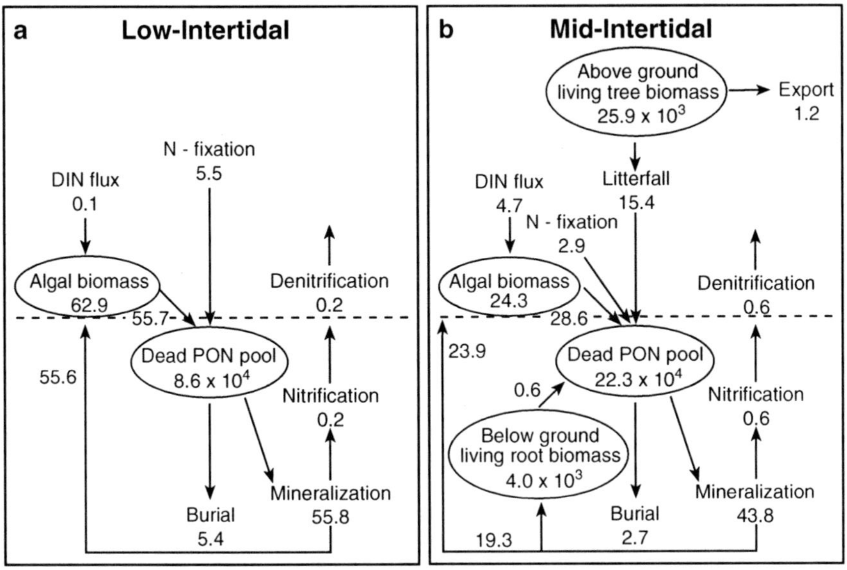

Fig. 5.8 Nitrogen budget for soils in low- and mid-intertidal *Rhizophora apiculata* forests on Phuket Island, Thailand (Data from Kristensen et al. 1995, 1998, 2000)

by the trees. Rates of ammonification are sufficient to sustain the rates of mangrove primary production. In the mixed mid-intertidal *Rhizophora* forests in Missionary Bay on Hinchinbrook Island, nitrogen cycling is equally rapid, sustaining high rates of mangrove primary production (Fig. 5.10). The pool of dissolved nitrogen is highly dynamic, with short turnover times implying rapid rates of uptake, presumably by the trees as there is little or no microalgal cover (Alongi 1996, 2005a). Most of the solute taken up at the soil–water interface appears to be used by roots. The burial rate of nitrogen is unknown, but is presumably a small proportion of nitrogen input, considering the fairly high rates of tree production in these forests.

Evidence for a close coupling between trees, microbes, and pore water nutrients in this forest can be gleaned from studies of DOC and DON dynamics (Stanley et al. 1987; Boto et al. 1989). Unless soils are poisoned to kill microbes and inhibit root uptake, significant release of DOC and DON from the soil is rarely detected, despite a strong concentration gradient. The non-protein amino acid, β-glutamic acid, is a major component of the interstitial N pool, yet is immobilized unless soils are poisoned. Once poisoned, large amounts of the amino acid diffuse into the overlying tidal water. Transport of DOC and DON is so rapid between trees, microbes, and the soil, that there is little, if any, residual pool of dissolved organic matter.

Organic nitrogen derived from roots supports high rates of ammonification in mangrove soils (Nedwell et al. 1994). Ammonification is the first key step in the nitrogen cycle, whereby nitrogen principally in the form of proteins and nucleotides are hydrolyzed and catabolized by ammonifying bacteria to be liberated as

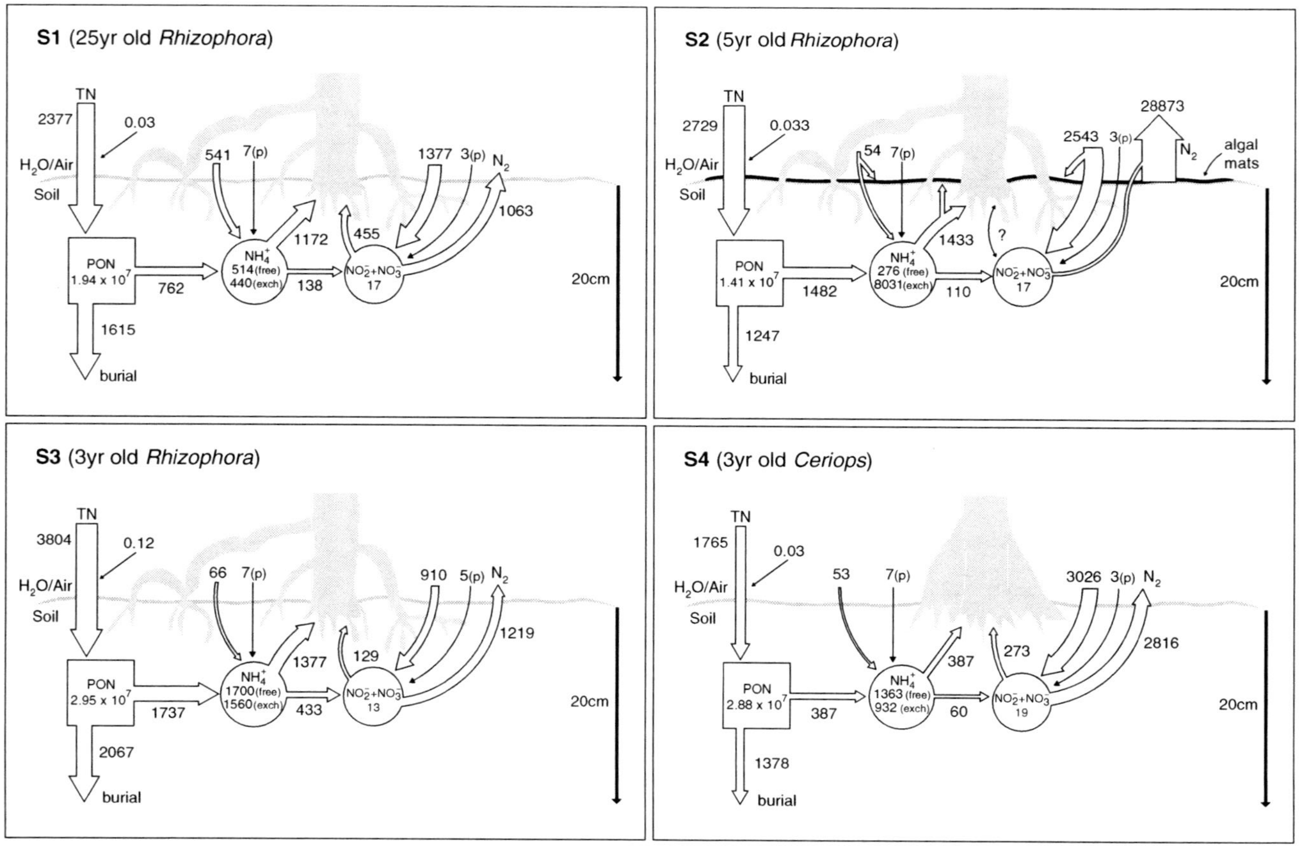

Fig. 5.9 Nitrogen budgets for four different mangrove forests in Sawi Bay, Thailand (Updated from Alongi et al. 2002)

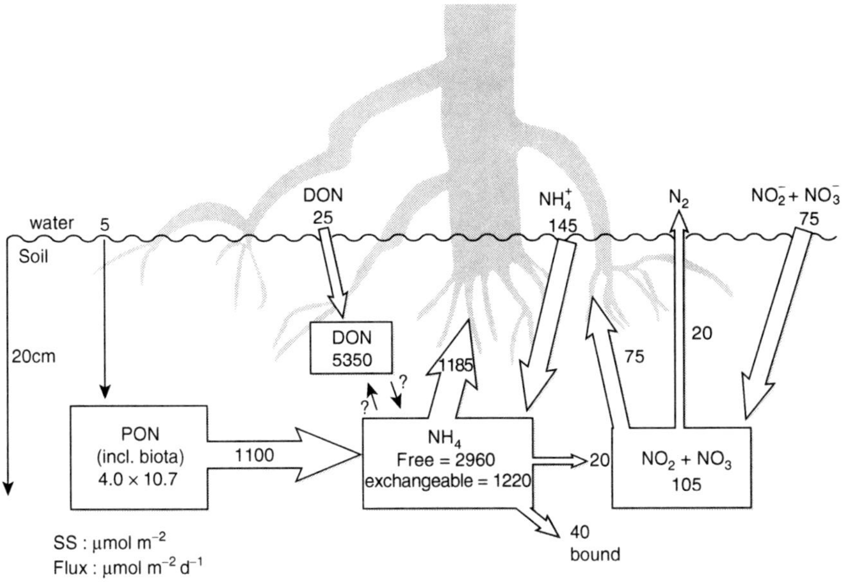

Fig. 5.10 Nitrogen budget for the mature mixed *Rhizophora* forests in Missionary Bay, Hinchinbrook Island (Updated from Alongi et al. 1992)

NH_4^+. Ammonification is not easily or accurately measured in saline soils, particularly those containing substantial amounts of dead and live fine roots. Using ^{15}N, Nedwell et al. (1994) measured rates of ammonification of 6.5–21.8 mmol m^{-2} day^{-1} over a soil depth of 8 cm with a very rapid turnover on the order of a few hours. Ammonification of a similar order of magnitude was estimated in mangrove soils of Joyuda Lagoon on the west coast of Puerto Rico (Morell and Corredor 1993) and in mangroves in Selangor, Malaysia (Shaiful et al. 1986).

Whether or not rates of ammonification are rapid is relative, depending on rates of total nitrogen input to the soils and rates of tree production. As the rate of organic nitrogen input to the soil increases, the rate at which this organic nitrogen is mineralized increases proportionally, as shown in Fig. 5.11 for mangrove soils in forests of China, Malaysia, and Thailand. Rates of ammonification in mangrove soils on Phuket Island (500–1,540 µmol m^{-2} day^{-1}) were similar to those measured in Sawi Bay (500–2,260 µmol m^{-2} day^{-1}), suggesting rates of nitrogen input similar to those measured in Sawi Bay (3–8 kg soil m^{-2} year^{-1}; Alongi et al. 2001).The mineralization and burial efficiencies did not correlate with rate of nitrogen input, but the mineralization efficiencies ranged narrowly from 67–92%. Burial efficiencies exhibited a wider range of 4–31%, but burial was <15% at eight of the ten forests. For these Asian forests, the vast bulk of nitrogen is mineralized and taken up rapidly in proportion to input, with little N being buried in the soil, having considerable impact on nitrogen retention in the forest; other means of retaining N must therefore be utilized to conserve N.

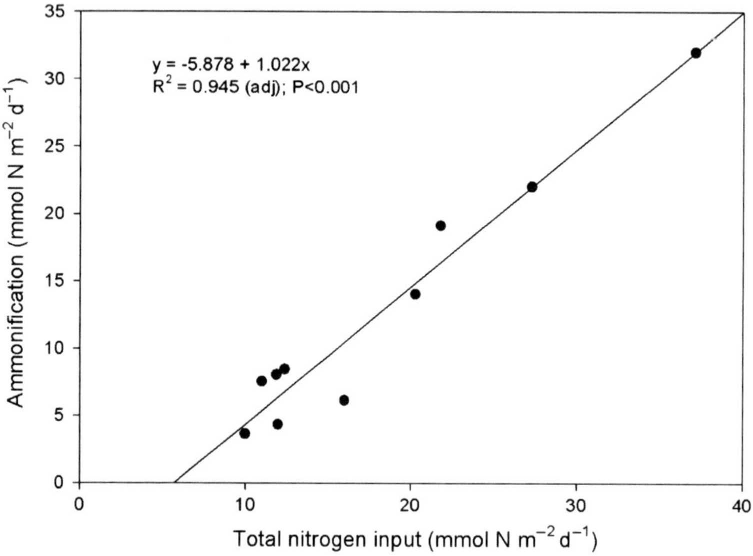

Fig. 5.11 The relationship between ammonification rate and rate of total input of nitrogen to soils in mangrove forests of Thailand, China, and Malaysia (Data from Alongi et al. 2002, 2004a, 2005b)

Denitrification is another important pathway of nitrogen loss affecting forest growth. Collating the available data (Iizumi 1986; Morell and Corredor 1993; Nedwell et al. 1994; Rivera-Monroy et al. 1995b; Rivera-Monroy and Twilley 1996; Kristensen et al. 1998; Alongi et al. 1999, 2000a, b, 2001, 2002, 2004a, 2005b, 2008; Joye and Lee 2004; Lee and Joye 2006), 81 sets of measurements show that N_2 losses to the atmosphere range from 0 to 11,000 μmol N m^{-2} day^{-1} yielding a mean denitrification rate of 1,532 μmol N m^{-2} day^{-1} with a standard error of 281 μmol N m^{-2} day^{-1} and a median of 226 μmol N m^{-2} day^{-1}. As with nearly all bacterial processes, measured values are highly method-dependent, so these values must be considered cautiously.

Denitrification is regulated by nitrate availability, temperature, salinity, and soil organic matter content. Nitrate concentration is the prime regulatory factor, as increasing the supply of nitrate increases the rate of denitrification in mangrove soils (Rivera-Monroy et al. 1995b; Rivera-Monroy and Twilley 1996; Joye and Lee 2004; Lee and Joye 2006). The lowest and highest rates of denitrification were measured in the Matang Forest Reserve in Malaysia with rates ranging from zero to 11,000 μmol N m^{-2} day^{-1}. The lowest rate was measured in a 5-year old forest and the highest rate was measured in an 85-year old stand of *Rhizophora apiculata*. It is unclear why these different aged forests exhibited such disparate rates other than to note that the nitrogen cycle is less disturbed and has been operating in the old growth forest for a longer period of time. A plot of the relationship of soil denitrification versus age of these *Rhizophora apiculata* forests (Fig. 5.12) shows a significant positive relationship, but the regression is skewed toward the

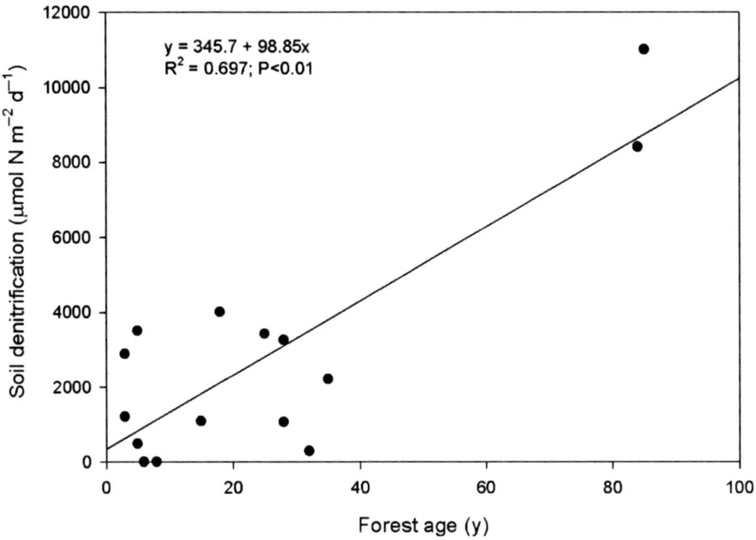

Fig. 5.12 The relationship between rates of soil denitrification and age of *Rhizophora apiculata* forests in Southeast Asia (Data from Alongi et al. 2000a, b, 2001, 2002, 2004a, 2008)

measurements from the oldest forest. The data does not suggest a relationship within the forest age span of 3–35 years. Denitrification did not relate either to rates of ammonification or rates of nitrogen input.

Factors other than nitrite availability therefore come into play in regulating the rate of denitrification. The presence of benthic microalgal mats, for instance, stimulates denitrification (Joye and Lee 2004; Lee and Joye 2006). On Twin Cays off Belize, benthic mats composed of filamentous, heterocystous and coccoidal cyanobacteria, purple sulfur bacteria, and heterotrophic bacteria, are dense within dwarf mangrove stands, playing an important role in nitrogen cycling. In these mats, nitrate is an important limiting factor for denitrification, and nitrogen fixation is regulated mainly by the sensitivity of the nitrogenase enzyme to oxygen inhibition. The size and thus the overall contribution of the mats to ecosystem nitrogen cycling is controlled by the seasonal and tidal frequency of wetting as well as elevation. Similar regulatory factors were discerned in mangrove soils of Terminos Lagoon in Mexico (Rivera-Monroy et al. 1995b; Rivera-Monroy and Twilley 1996), where nitrate availability was shown to be the major controlling factor. However, denitrification was not linked to nitrification, and uptake from tidal waters was not the principal source of nitrate as ^{15}N experiments suggested that nitrogen was retained in the soil. In Thai mangroves, Alongi et al. (2002) found that denitrification was highly variable and uncoupled to nitrification, similarly implying that nitrogen is immobilized in the soil.

The data from Thailand suggests that extensive uptake of ammonium by mangroves is associated with low rates of nitrification, which may be inhibited by soluble tannins or anoxic metabolites, such as sulfides. In contrast, Kristensen et al. (1998) found that nitrification contributed 90% of the nitrate needed for denitrification in mangrove soils on Phuket Island. They provide among the very few reliable measurements of nitrification in mangrove soils, as rates measured using older, unreliable methods indicate low rates of ≈1–2 μmol m^{-2} day^{-1} in mangrove soil (Iizumi 1986; Shaiful et al. 1986; Shaiful 1987). In the Phuket mangroves, Kristensen et al. (1998) measured rates of 12–43 μmol m^{-2} day^{-1}. Using stable isotopes, Rivera-Monroy and Twilley (1996) measured initial nitrification rates of 672 μmol m^{-2} day^{-1} in experimental cores taken from Mexican mangroves, with rates declining below detection within a few days owing to $^{15}NH_4^+$ immobilization in the soil. Differences in nitrification rates among locations are more likely to be the result of methodological shortcomings than of real differences among forests.

On average, about 15% (range: 3–47%) of total nitrogen input into mangrove soils is denitrified. In other estuarine and marine deposits, the percentage of N lost via denitrification is normally within the range of 15–70% (Seitzinger 1988). It thus appears that denitrification as an export pathway of nitrogen is of proportionally less significance in mangroves than in other aquatic ecosystems. Factors that play a role in limiting losses via denitrification include microbial conversion of inorganic N into organic forms, low rates of nitrification, low nitrate availability, high C:N of the source material, and inhibition due to the presence of toxic metabolites. These may indeed be the processes (as suggested earlier) by which N is retained within forests.

The anaerobic conversion of NO_2^- to N_2, a process known as anaerobic ammonium oxidation or 'anammox', first discovered in temperate marine deposits, has been detected in mangrove soils (Meyer et al. 2005). Along a subtropical mangrove river, Meyer et al. (2005) found that the potential rate of anammox increased up river, correlating with nitrite production and the size of the nitrite pool in the soil. Although NO_2^- accumulates from both nitrification and nitrate plus nitrite reduction, the latter process regulates nitrite availability in suboxic soil layers. Denitrification also provides a substrate for the anammox process. Assuming that anammox is an important transformation process in other mangrove soils, the denitrification rates obtained by Alongi and his colleagues (Alongi et al. 1999, 2000b, 2001, 2002, 2004a, 2005b, 2008) may be overestimates. This is because some of the N_2 measured as part of the direct gas procedure used may have been the result of anaerobic NH_4^+ oxidation rather than activities of denitrifying bacteria.

Denitrification is often thought to be counterbalanced by nitrogen fixation, the process by which diazotrophic prokaryotes transform atmospheric N_2 into ammonia via nitrogenase activity. However, rates of nitrogen fixation are lower than denitrification rates in most benthic systems, and such appears to be the case in mangrove soils where relatively low nitrogen fixation is mediated by sulfate reducers and microbial assemblages associated with the rhizosphere (Alongi 2005a). Collating the available data (Iizumi 1986; Morell and Corredor 1993; Nedwell et al. 1994; Kristensen et al. 1998; Alongi et al. 1992, 1999, 2000b, 2001, 2002, 2004a, 2005b, 2008; Joye and Lee 2004; Lee and Joye 2006), nitrogen fixation ranges from zero to 4,316 μmol N m^{-2} day^{-1} with a mean nitrogen fixation rate of 616 μmol N m^{-2} day^{-1}, a standard error of 145 μmol N m^{-2} day^{-1}, and a median of 18 μmol N m^{-2} day^{-1} (assuming a 4:1 ratio of acetylene to N_2 reduction). These rates are less than those for denitrification indicating that nitrogen fixation does not offset denitrification losses. These rates are also less than those measured in salt marshes and seagrass beds (Howarth et al. 1988). Most rates measured in mangrove forests, however, pertain to activity at the soil surface; nitrogen fixers are very active elsewhere in the canopy. Nitrogen fixation has been detected on prop roots, litter, fresh leaves, bark, logs, and other wood in the forest (Alongi et al. 1992). Further studies are needed to determine whether or not nitrogen fixation on these forest wood components equals losses via denitrification and anaerobic ammonium oxidation.

High rates of nitrogen fixation may also be occurring deep within the extensive root systems of mangroves. These N_2-fixers are in a mutualistic relationship with non-N_2 fixing bacteria and the trees, and provide the bulk of nitrogen for immediate plant use (Sengupta and Chaudhuri 1991; Holguin and Bashan 1996; Bashan et al. 1998; Rojas et al. 2001; Ravikumar et al. 2004; Naidoo et al. 2008). In culture experiments, nitrogen fixation by the bacterium *Azospirillum brasilense* is enhanced when cultured in the presence of a strain of a mangrove rhizobacterium *Staphylococcus* sp. suggesting metabolic by-products produced by the rhizobacteria benefit the growth of the nitrogen fixers (Holguin and Bashan 1996). Transfer of fixed nitrogen to mangrove roots from filamentous cyanobacteria similarly enhances the growth of mangrove seedlings (Bashan et al. 1998). Indeed, nitrogen-fixers that reside in the rhizosphere are now being used to promote mangrove growth for forest restoration (Ravikumar et al. 2004). Such results may help to explain the patterns of nitrogen fixation in young versus mature mangrove forests in India (Sengupta and Chaudhuri 1991). In this study, nitrogen fixation rates peaked in rhizospheres excised from seven mangrove species representing early pioneers in the Ganges. Rates of nitrogen fixation declined in roots excised from species representing late-successional stages, implying that in the early stages of development, mangroves receive a significant natural boost from nitrogen-fixing bacteria.

Nitrous oxide (N_2O) is an intermediate product of nitrification and denitrification and is an important gas, being involved in the greenhouse effect with a global warming potential 200–300 times that of CO_2 and participating in ozone dynamics in the upper atmosphere (Canfield et al. 2005). Only a few studies have measured N_2O fluxes from mangrove soils (Table 5.7). Rates vary from undetectable or net

Table 5.7 Rates of nitrous oxide flux (μmol m^{-2} day^{-1}) across the mangrove soil–air interface in forests worldwide

Location	Forest type	N_2O	Reference
Puerto Rico	*A. germinans*	2.9–7.9	Corredor et al. (1999), Bauza (2007)
Puerto Rico	*R. mangle* (bird rookery)	186.7	Corredor et al. (1999), Bauza (2007)
Puerto Rico	*R. mangle* (untreated)	1.0–16.1	Muñoz-Hincapié et al. (2002), Bauza et al. (2002)
China	*K. candel*	0–106.5	Alongi et al. (2005b)
Australia	*A. marina*	−2.2 to 35.5	Allen et al. (2007)
India	*R. apiculata, A. marina*	0.5–28.8	Barnes et al. (2006, 2007), Upstill-Goddard et al. (2007)
Vietnam	*K. candel*	−22.6 to 330.0	Imamura et al. (2007)
Indonesia	*R. apiculata*	0	Alongi et al. (2008)

uptake, to release rates as high as 330 μmol m^{-2} day^{-1}. Studies in Puerto Rico offer some insight into factors controlling rates of N_2O flux. With addition of either NH_4^+ or NO_3^-, rates of N_2O efflux from soils increased dramatically during field experiments (Fig. 5.13) with no sign of saturation with added nitrate (Muñoz-Hincapié et al. 2002). Also found was a diel cycle with peak emission rates centering on peak daylight; nitrous oxide emissions were attributed to nitrification and nitrate availability (Bauza et al. 2002; Bauza 2007). A similar enhancement effect was found by Meyer et al. (2008) using microsensors. They found that N_2O emissions were linked to both nitrification and denitrification. Factors that affect nitrate availability

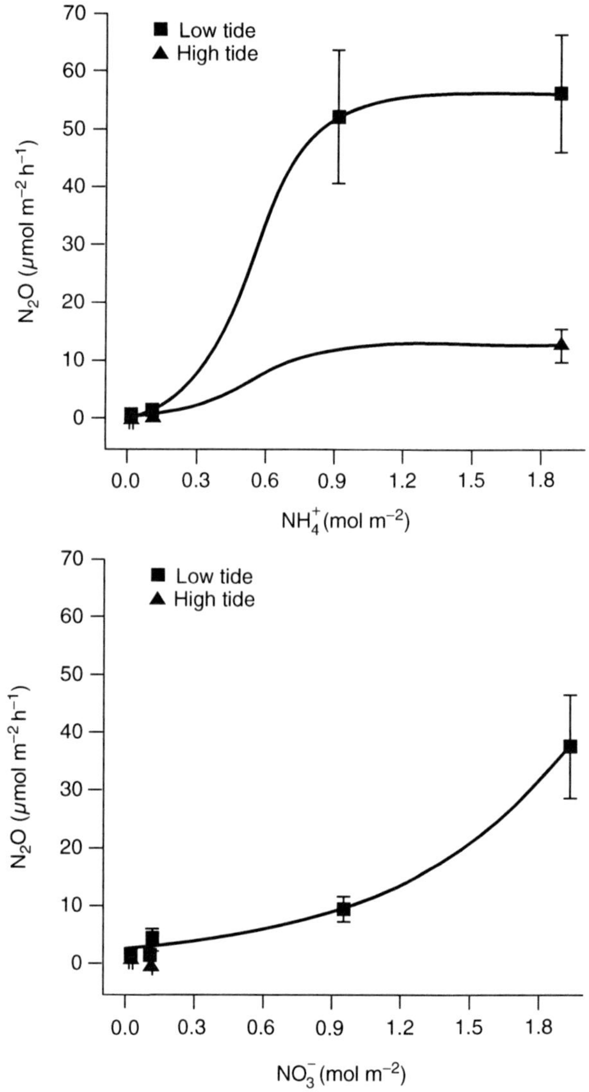

Fig. 5.13 Nitrous oxide fluxes in response to ammonium and nitrate addition to mangrove soil during low tide in the dry season (Modified from Muñoz-Hincapié et al. 2002)

(e.g., nitrogen loading, redox status, and temperature) therefore also impact on emission rates of N_2O (Kreuzwieser et al. 2003; Upstill-Goddard et al. 2007).

The studies of Upstill-Goddard and his colleagues in India have demonstrated not only that mangrove waters are important emission sites for CO_2 and CH_4 (see Section 6.2.1) but also for N_2O (Upstill-Goddard et al. 2007; Barnes et al. 2007). Using either floating chambers or applying a transfer velocity-wind speed relationship to N_2O concentration differences between water and air, they measured rates ranging from 2.88–31.2 μmol m^{-2} day^{-1}. More significantly, emission rates were greater from creek water than from mangrove soils. At these same sites, concentrations of nitrous oxide in the water-column corresponded to the tidally-induced changes in dissolved inorganic nitrogen concentrations (Fig. 5.14) with peak levels at low tide and minimum concentrations at high tide. This pattern was interpreted as being consistent with tidal pumping in which high porewater concentrations of nutrients and dissolved gases seep into the creek waters from adjacent mangrove soils after the gradual release of hydrostatic pressure towards low tide (Barnes et al. 2006, 2007). In Vietnam and Japan, Imamura et al. (2007) similarly measured net release of N_2O from tidal creeks on the order of 4–72 μmol m^{-2} day^{-1} in Vietnam and −50 to 150 μmol m^{-2} day^{-1} in Japanese mangroves. Considering that the area of adjacent tidal water is greater than the forest area in most coastal mangrove ecosystems, tidal waters must be considered significant sources of atmospheric N_2O (and CO_2 and CH_4) at the ecosystem-scale.

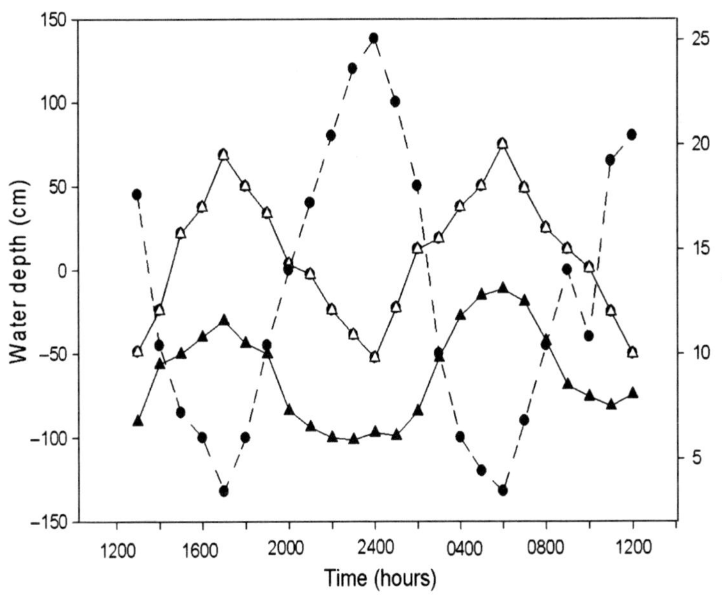

Fig. 5.14 Changes in water-column concentrations of N_2O (white triangles) and dissolved inorganic nitrogen (black triangles) over a tidal cycle (black dots with dotted line) in the dry season at the mangroves of Wright Myo, Andaman Islands, India (Modified from Barnes et al. 2006, 2007)

5.4.6 Aspects of Phosphorus Cycling

Our knowledge of phosphorus cycling in mangrove soils has not advanced noticeably since early last decade (Alongi et al. 1992). Many measurements have been made of the flux of dissolved inorganic phosphorus across the soil–water interface and a few studies have assessed the extent of P limitation on forest growth, but these measurements tell us little about the rates and pathways of phosphorus cycling within the soils and how and under what conditions this element is transferred to the trees.

We do know that during the decomposition of organic matter, organic phosphorus is partly assimilated and partly released as dissolved inorganic phosphorus (DIP) by microheterotrophs, with the degree of partitioning primarily dependent on the C:P ratio of the source material (Canfield et al. 2005). DIP is available as either $H_2PO_4^-$ or HPO_4^{2-} depending on soil pH, but readily forms insoluble precipitates with a number of divalent cations, particularly Ca^{2+} and Fe^{2+}. Under oxic conditions, phosphate also adsorbs onto positively charged clay surfaces and oxides of Fe^{3+} and Al^{3+}. Under anoxic conditions, Fe^{3+} is reduced to Fe^{2+} and other ferrous minerals (e.g., siderite) that are less adsorptive; P is subsequently released into the interstitial water. The P cycle is thus closely associated with metal and redox chemistry of the soil.

Mangrove soils, due to their normally high organic content compared with unvegetated saline deposits, may contain a high proportion of organic-bound P, up to 75–80% of the total extractable P, in some locations (Alongi et al. 1992). Recent work has suggested quite variable proportions between organic and inorganic fractions in relation to grain size and origin (terrestrial versus marine) and stage of forest development, with many soils exceeding 50% of P within the inorganic fractions (Fabre et al. 1999; Koch et al. 2001; Chambers and Pederson, 2006; Lai and Lam 2008). In Micronesian forests, soluble reactive phosphorus concentrations relate to redox conditions and species composition of the forests, the latter relationship engendered either by the trees or to micro-scale differences among stands composed of different species (Gleason et al. 2003).

The inorganic fractions of the total P pool have attracted the most attention because inorganic forms represent the largest potential pool available for use by the trees. The phosphorus cycle is relatively simple in that there are no gaseous phases, but the relationship between microbial activities and changes in P geochemistry can be highly complex and difficult to measure. Transformation of P can be categorized as: (1) abiotic (precipitation, dissolution, desorption, adsorption, chemisorption) and (2) biotic (assimilation, excretion, hydrolysis). Organisms participate in the cycle by excreting soluble reactive phosphorus, mineralizing organic phosphates, and in the case of autotrophs, taking up dissolved inorganic P. The availability of P, however, is not regulated by biological processes, but by geochemical reactions. "Readily available" P is rapidly incorporated onto clay particles and metal oxyhydroxides, and immobilized by precipitation as Ca, Fe, Al salts, limiting the available P pool for organisms. Organic phosphates, mainly phosphate esters originating from living cell tissue, are often resistant to hydrolysis and therefore limiting to microbes and plants. Early (Hesse 1962, 1963; Boto and Wellington 1983) and more recent (Tam and Wong 1996; Holmboe et al. 2001) experiments show that when phosphate is added to mangrove soils, it is quickly immobilized onto Fe and Al oxyhydroxides and into easily exchangeable fractions. Mangrove soils with such a high adsorptive capacity function as phosphorus sinks.

As phosphorus is closely tied to the iron and sulfur cycles, it is not surprising that the distribution of mangroves has been linked to concentrations of phosphorus and to the extent of sulfate depletion, as found by Sherman et al. (1998) in *Laguncularia*-dominated forests of the Dominican Republic. These data imply that processes that affect the availability of iron and sulfur in turn affect the availability of phosphorus. That is, we can speculate that with greater rates of sulfate and iron reduction, more Fe oxyhydroxides are reduced to Fe^{2+} resulting in more iron-bound P being released into the interstitial water pool and being available for uptake.

The uptake of soluble P by mangroves closely involves mutualistic interrelationships among bacteria, fungi, and tree roots. Arbuscular mycorrhizal fungi in the mangrove rhizosphere benefit from oxygen translocated by the trees to their roots, and the presence of vesicles (nutrient storage organs) in the root cells of some mangrove species (Kothamasi et al. 2006) suggests that fungal symbionts play a role in nutrient uptake; phosphate-solubilizing bacteria associated with the roots and fungi may release phosphate that could be taken up by the fungal hyphae and transferred to the host or taken up directly by the roots. Phosphate-solubilizing bacteria have been found in the roots of a number of mangrove species (Vazquez et al. 2000; Rojas et al. 2001; Kothamasi et al. 2006; Bashan and Holguin 2002) and their presence has been shown to increase rates of other bacterial processes, such as nitrogen fixation (Rajas et al. 2001). How these bacteria solubilize phosphate is unclear, but culture experiments suggest that organic acids produced by the bacteria may dissolve calcium phosphate (Vazquez et al. 2000). As suggested by Bashan and Holguin (2002), such plant growth-promoting bacteria and fungi can be used as a tool for reforestation. In any event, further research is needed to obtain a holistic picture of the role of phosphorus in mangrove ecosystems.

Chapter 6
Ecosystem Dynamics

6.1 Introduction

An ecosystem, as defined by Odum and Barrett (2005) is a functional unit that "includes all the organisms in a given area interacting with the physical environment so that flows of energy and materials leads to clearly defined biotic entities and cycling of materials between living and nonliving components". It is the first complete ecological unit from which we can metaphorically see the forest for the trees. From this perspective, the carbon balance of mangrove ecosystems and the cycling of nitrogen and mineral elements in relation to ecosystem development (that is, young versus old ecosystems) will be assessed, including how the delicate balance of these cycles are (or aren't) altered by human disturbance. Finally, some holistic models of mangrove food web dynamics will be examined before moving on to how an energetics perspective can help delimit the key conservation and management issues of sustainability and ecosystem services.

6.2 Material Exchange: The Outwelling Concept

Tides represent a form of energy subsidy for mangrove ecosystems, in that tides (and to a lesser extent, waves) do the work of transporting particulate and dissolved materials, gases, and other by-products between the forest and the adjacent coastal zone (see Chapter 3). This function is a corollary of source-sink energetics in which excess organic production by mangroves (the source) is exported to the less productive, coastal nearshore zone (the sink). In reality, of course, energy and materials are also imported into mangrove ecosystems by the same tides. The idea that the fertility of estuarine wetlands may contribute nutrients to sustain productivity in the adjacent nearshore came from a brief commentary by Eugene Odum (1968) in which he suggested that high productivity in coastal areas results either from 'upwelling' of deep water or from 'outwelling' of nutrient and organic detritus from fertile hot spots such as salt marshes, reefs, and macrophyte beds (Odum 2000). The 'outwelling' hypothesis generated a great deal of subsequent research effort on the exchange of particulate and dissolved nutrients between wetlands and adjacent coastal waters, including a number of research programs on whether or not mangroves outwell substantial amounts of material to the tropical coastal zone.

After nearly 3 decades of research, we now know that the amount of organic detritus potentially available for export from mangroves is influenced by many factors, including net forest primary production, tidal range, ratio of mangrove to watershed area, lateral trapping, high salinity plugs, total mangrove area, frequency of storms, amount of rainfall, volume of water exchange, and the extent of activities of crabs and other litter-feeding fauna (Twilley 1988). The number of factors and their nature is such that each system is unique; some mangroves export nutrients and some do not.

6.2.1 Carbon Export to the Coastal Ocean and the Atmosphere

Most data on material exchange involve estimates of the export of particulate organic carbon, mainly as litter, from mangrove estuaries. Updating the carbon export data in Table 6.1 from Jennerjahn and Ittekkot (2002) gives a revised annual carbon export estimate of 15.3 mol C m^{-2} (Table 6.1). Assuming that the total mangrove area worldwide is 15,763,000 ha (FAO 2003), multiplying both numbers gives a mean value of 29 Tg $year^{-1}$. This estimate is at the low end of the range (30–50 Tg $year^{-1}$) first calculated by Twilley et al. (1992) and less than the 46 Tg $year^{-1}$ estimated by Jennerjahn and Ittekkot (2002). Further assuming a mean above-ground net primary productivity of 44.5 mol C m^{-2} $year^{-1}$ (Section 2.5.3), the export of POC equates to 32%, or roughly one-third, of total NPP. Obviously, the amount of export will vary greatly among estuaries depending on the relative importance of the factors listed above, but these estimates indicate that mangrove POC export

Table 6.1 Estimates of export of particulate organic carbon (mol C m^{-2} $year^{-1}$) from mangrove estuaries worldwide

Location	Export	Reference
Rookery Bay, Florida	5.3	Twilley (1985b)
South Florida	15.5	Twilley (1985b)
Tuff Crater, New Zealand	9.3	Woodroffe et al. (1985a, b)
Darwin Harbour, Australia	26.7	Woodroffe et al. (1988), Burford et al. (2008)
Matang, Malaysia	19.1	Gong and Ong (1990), Alongi et al. (2004a)
Klong Ngao, Thailand	0.1	Wattayakorn et al. (1990)
Itacuruca, Brazil	18.3	Lacerda (1992)
Fly River, Papua New Guinea	23.8	Robertson and Alongi (1995)
Missionary Bay, Australia	27.7	Alongi (1998)
Hinchinbrook Channel, Australia	10.4	Ayukai et al. (1998)
Sawi Bay, Thailand	5.9	Alongi et al. (2000c)
Caeté estuary, Brazil	16.1	Dittmar et al. (2001)

could account for 10–11% of the total input of terrestrial carbon into the ocean and 12–15% of the total carbon accumulation in sediments on the continental margin. Similarly, Dittmar et al. (2006) estimate that about 10% of terrestrially-derived DOC exported to the ocean is contributed by mangroves. Given their small area relative to other ecosystems, mangroves are thus contributing disproportionately to POC flux to the coastal ocean.

Naturally, the largest exports come from macrotidal and mesotidal estuaries (e.g., Darwin Harbour, Missionary Bay, and the Fly River), and some of the smallest exports come from microtidal systems (e.g., Sawi Bay), underscoring the importance of tidal regime and the fact that ebb tides tend to be stronger than flood tides. Other studies have measured net export, but either the amount of material was unquantified or difficult to determine from the information provided (Hemminga et al. 1994; Harrison et al. 1997; Rivera-Monroy et al. 1998; Ovalle et al. 1999; Davis et al. 2001; Pradeep Ram et al. 2003).

Clearly, most mangroves export POC, but the patterns of both POC and DOC exchange often differ within the same estuary with change in season. For example, in Missionary Bay in northern Australia, the Coral Creek mangroves annually export, on average, 27.7 mol POC m^{-2}; DOC exchange varies seasonally with net import during the summer and a slight import overall of 0.6 mol DOC m^{-2} $year^{-1}$ (Robertson et al. 1992).

In other systems, such as in basin forests or in microtidal systems, a greater proportion of material is exported in dissolved form (southwest Florida: Twilley 1985b, Sawi Bay, Thailand: Alongi et al. 2000c). In the only shelf-scale study of mangrove export of DOC, Dittmar et al. (2006) found greater outwelling of DOC (12 mol C m^{-2} $year^{-1}$) than did earlier small-scale studies in the same region off Brazil (4 mol C m^{-2} $year^{-1}$; Dittmar et al. 2001). The difference was attributed to the fact that small-scale studies usually do not account for the gradual release of DOC from detritus suspended or floating in offshore waters (Kristensen et al. 2008). Off Brazil, Dittmar and his colleagues (Dittmar et al. 2001, 2006; Dittmar and Lara 2001a, b; Schories et al. 2003) found that 13 mol C m^{-2} $year^{-1}$ is exported as detritus from tidal creeks and well-developed forests, with an additional 3 mol C m^{-2} $year^{-1}$ exported as smaller particles and 4 mol C m^{-2} $year^{-1}$ exported as DOC; the combined export equates to about 40% of total litterfall. Of the total export of 20 mol C m^{-2} $year^{-1}$, roughly 60% is eventually transported further offshore in the form of DOC derived from further reworking of particulate matter across the continental shelf. From these data we can estimate a global DOC export from mangroves of 14 Tg $year^{-1}$.

DOC exported from mangrove estuaries has a unique chemical signature, derived from degradation products of mangrove detritus incorporated into the soil and often seeping out of the porewater (Section 5.2); the origin of this DOC is often deduced from a clear tidal signature, with mangrove DOC leaving the estuary during the ebb tide and DOC of marine origin entering the estuary during the flood tide (Bouillon et al. 2003, 2007b, c). Thus, most of the DOC exported does so by way of tidally-induced porewater flow (Schories et al. 2003).

Most of the DOC is refractory to immediate microbial, physical, and photochemical decay. In a detailed study of the decomposition of plant-derived

DOM along a seaward gradient in the Florida Everglades, Scully et al. (2004) observed that polyphenolic compounds are degraded mainly via photolysis and that high molecular weight compounds are degraded primarily through microbial and physicochemical processes. These latter processes initiate the formation of refractory, highly colored, high molecular weight polymers. Thus, there is some rapid degradation of mangrove-derived DOC followed by much slower transformation of DOM. This finding is similar to that found by Maie et al. (2008) for DON compounds in mangrove and coastal waterways. A high proportion of tannins precipitate upon exposure to salt and also by sorption to sediments in the estuary. Some DON co-precipitates with the tannins, and these complexes are highly reactive, with a half-life in the water-column of <1 day. Proteins are released gradually from these DON-tannin mixtures, so tannins play an important role in retaining nitrogen in the system, buffering the loss of N through the prevention of rapid tidal export (see Section 4.2).

Although there is a clear pattern of export from mangroves, what role this material has as a nutrient subsidy for offshore food webs is unclear. A picture is emerging, however, which indicates that the extent of mangrove influence is normally restricted to a few kilometers offshore (Lee, 1995; Alongi 1998; Baltzer et al. 2004) where seagrass beds and coral reefs can be supplemented by DIC derived from mangrove respiration (Jennerjahn and Ittekkot 1997; Ovalle et al. 1999; Machiwa 2000; Machiwa and Hallberg 2002; Mfilinge et al. 2005; Bouillon et al. 2008). DOC is often transported further offshore than particulate material, especially from mangroves lining large deltas such as the Amazon (Dittmar et al. 2001). This material, in fact, has been chemically traced to the edge of the continental margin.

The limited impact of mangrove detritus on offshore food webs can be attributed to several factors:

- Local geomorphology and hydrodynamics of mangrove estuaries mitigates against extensive outwelling of labile material.
- The presence of coastal boundary zones off tropical coasts or a high salinity plug in the dry season can efficiently trap litter and suspended particles inshore.
- Most material lost from the system is either highly refractory particulate matter or DOC, the labile fraction of which can be considerably degraded further in the water-column.

These generalizations do not hold so well for large river systems such as the Amazon and Indus.

Recent studies of water-air CO_2 fluxes from mangrove waterways and adjacent inshore waters suggest that pelagic mineralization of organic matter and subsequent emission of CO_2 to the atmosphere could represent another significant pathway of carbon export from mangrove ecosystems (Ghosh et al. 1987; Richey et al. 2002; Borges et al. 2003; Bouillon et al. 2003, 2007a–c; Biswas et al. 2004; Barnes et al. 2006; Ramesh et al. 2007; Upstill-Goddard et al. 2007; Koné and Borges 2008; Ralison et al. 2008). These studies consistently found that mangrove waters are oversaturated in CO_2 as a direct result of pelagic respiration and CO_2 respired within the forest soils that is dissolved in the porewater and transported laterally by tidal pumping to the adjacent creeks and waterways (Borges et al. 2005). Flux rates, measured either using flux chambers or by modelling air–sea exchange, vary greatly with tidal stage, temperature, precipitation, and location. The Sunderbans bordering the Bay of Bengal, where mangroves constitute nearly 3% of the world's total mangrove area, contributes greatly to the control of CO_2 exchange between air and sea. Measuring diurnal and seasonal variations of CO_2 exchange, Biswas et al. (2004) found that in the post-monsoon months CO_2 saturation and fluxes are minimal and maximal in the pre-monsoon and early monsoon months; the waters of the Sunderbans are heterotrophic throughout the year. The Sunderbans mangrove forest emits 314.6 µmol C m^{-2} day^{-1} of CO_2 to the atmosphere even though nearly 60% of CO_2 emitted by the ecosystem is removed from the atmosphere by biological processes (e.g., plant uptake).

Averaging the data in Table 6.2 gives a mean flux rate of 43.3 mmol C m^{-2} day^{-1}. Using a higher mean rate of 72 mmol C m^{-2} day^{-1}, Koné and Borges (2008) estimated that CO_2 emission from mangrove waters corresponds to ≈7% of the total emission from oceanic waters at subtropical and tropical latitudes, and about 24% of the total CO_2 emissions from coastal waters globally. The percentage contribution by mangrove waters to global emissions will change as more measurements are made, but these preliminary estimates indicate that, like organic carbon, the mangrove contribution of inorganic carbon to the global ocean is disproportionate to their relatively small area.

Table 6.2 Rates of water–air exchange of CO_2 (mmol C m^{-2} day^{-1}) from mangrove waters

Location	Flux	Reference
Saptamukhi Creek, India	56.7 ± 37.4	Ghosh et al. 1987
Mooringanga Creek, India	23.2 ± 10.1	Ghosh et al. 1987
Itacuraçá Creek, Brazil	113.5 ± 104.4	Ovalle et al. 1990, Borges et al. 2003
Florida Bay, USA	4.6 ± 5.4	Millero et al. 2001
Mekong, Vietnam	42.1	Richey et al. 2002
Amazonas, Brazil	175.2	Richey et al. 2002
Nagada Creek, Papua New Guinea	43.6 ± 33.2	Borges et al. 2003
Gaderu Creek, India	56.0 ± 100.9	Borges et al. 2003
Norman's Pond, Bahamas	13.8 ± 8.3	Borges et al. 2003
Godavari, India	21.9 ± 26.1	Bouillon et al. 2003
Tidal Creeks, Godavari	70.2 ± 127.0	Bouillon et al. 2003
Kakinada Bay, India	8.3 ± 13.6	Bouillon et al. 2003
Sunderbans, India	3.2	Biswas et al. 2004
Ras Dege, Tanzania	33	Bouillon et al. 2007c
Adyar, India	17.8	Ramesh et al. 2007
Muthupet, India	31.8	Ramesh et al. 2007
Pichavarum, India	6.1	Ramesh et al. 2007
Shark River, Florida	43.8 ± 52.1	Koné and Borges 2008
Ca Mau, Vietnam	94.2 ± 50.9	Koné and Borges 2008
Betsiboka estuary, Madagascar	9.1 ± 14.2	Ralison et al. 2008

6.2.2 Dissolved Nitrogen and Phosphorus Exchange

The net direction of dissolved nutrient exchange between mangrove waterways and the adjacent coastal zone depends upon tidal range, extent of groundwater discharge (Section 3.3), ratio of evaporation to precipitation, rates of primary productivity, salinity, turbidity, pH, dissolved oxygen concentrations, and rates of microbial assimilation. Another driving force often overlooked is the extent to which porewater concentrations of nutrients exceed the demands of primary producers (Dittmar and Lara 2001c). Simply, a system will tend to export nutrients if there are more nutrients than needed for utilization within the ecosystem. Conversely, nutrients such as nitrogen will be imported into the system if there is not enough available. Anthropogenic changes sustained by estuaries may also lead to shifts in patterns of nutrient and material exchange. In the Red River estuary of Vietnam, where there has been a massive increase in mangrove production due to large-scale accumulation of sediments transported from upriver, the estuary has become a sink for N and P and this pattern is directly attributable to the increase in mangrove forests (Wösten et al. 2003).

Some nutrient species are imported while others can be exported from the same ecosystem (Table 6.3). For instance, Coral Creek in the Missionary Bay ecosystem, exports significant quantities of litter and some nitrate, but imports phosphate, silicate, ammonium, and DON (Boto and Wellington 1988; Alongi 1996). Some estuaries located in the wet tropics exhibit strong outwelling patterns (Ovalle et al. 1990; Ayukai et al. 2000; Mukhopadhyay et al. 2006); nearly all ecosystems exhibit seasonality of such patterns related to local weather, such as extended drought or intense storms. In short, there are no universal patterns of dissolved nutrient exchange between mangroves and adjacent coastal waters, even if the same ecosystem outwells particulate material.

Table 6.3 Net annual exchange of dissolved nutrients (mmol m^{-2} $year^{-1}$) for various mangrove estuaries worldwide. I = import, E =export

Location	NH_4	NO_3	PO_4	DON	Reference
Coral Creek, Australia	I	E	I	I	Boto and Wellington 1988
Sepetiba Bay, Brazil	E	E	E		Ovaille et al. 1990
Klong Ngao, Thailand		I	I		Wattayakorn et al. 1990
Estero Pargo, Mexico	I	I		E	Rivera-Monroy et al. 1995a
Conn Creek, Australia		E	E		Ayukai et al. 1998
Sawi Bay, Thailand	E	E	E		Ayukai et al. 2000
Taylor River, Florida	I	E			Davis et al. 2001
Bandon Bay, Thailand	E	E	E		Wattayakorn et al. 2001
Gazi Bay, Kenya	E	E	E		Mwashote and Jumba 2002
Okinawa, Japan	E	I			Kurosawa et al. 2003
Red River, Vietnam	I	I	I		Wösten et al. 2003

6.3 Carbon Balance in Mangrove Ecosystems

The balance between photosynthetic gains by plants and respiratory losses by all organisms is reflected in the exchange of carbon between the ecosystem, atmosphere, and the adjacent coastal ocean. This balance is called net ecosystem production (NEP) or net ecosystem exchange (NEE). NEP has become a crucial characteristic in ecosystem assessments of whether or not anthropogenic increases in atmospheric CO_2, resulting from fossil fuel combustion and the clearing of forests, has altered the carbon balance. Forests are particularly important storage sites for CO_2, reducing the impact of human inputs into the atmosphere (Perry et al. 2008).

Determining the carbon balance of a mangrove ecosystem is similar to that for an individual tree (see Section 2.5.2) in that carbon inputs from photosynthesis must be balanced by losses such as respiration and the shredding of leaves. At the ecosystem level, however, other inputs and outputs must be accounted for, such as groundwater, burial within the forest floor, tidal exchanges, river inputs, and respiratory losses from microbes, fauna, and flora. In disturbed ecosystems, human impacts also have to be accounted for, such as losses due to logging, clear felling, fishing, and gains from aquaculture and sewage. These inputs and outputs are needed to construct mass balance models to estimate net ecosystem production. But first, we will look at a more holistic approach to quantifying mangrove carbon balance.

6.3.1 Whole-Ecosystem Balances

Techniques borrowed from micrometeorology and terrestrial ecology have provided a new approach to measuring the exchange of CO_2 between forests and the atmosphere (Aber and Melillo 2001; Perry et al. 2008). One approach, the eddy covariance method, estimates the net exchange of CO_2 by measuring the vertical gradients of CO_2 from the forest floor to above the canopy (Fig. 6.1). With simultaneous measurements of wind speed, wind direction, and air temperature, the vertical CO_2 profile can be used to estimate carbon flux in and out of the forest over various time intervals.

The eddy covariance method captures the movement of air through the canopy which occurs as a rapidly oscillating set of eddies or irregular pockets of turbulent air shifting in response to the interaction between winds and the vortices induced by the presence of the foliage. Net carbon accumulation is occurring if the air moving out of the canopy has a lower CO_2 concentration than the air moving into the forest.

The first studies to use this method in mangroves were conducted by Monji and his colleagues (Monji et al. 1996, 2002a, b; Monji 2007) at Phangnga in southern Thailand. Using several different gradient analyses (due to problems measuring CO_2 concentrations during heavy rainfall and from both wet and dry soils), Monji et al. (2002a, 2000b) found that, as expected, there are clear diurnal changes in CO_2 flux above the canopy (Fig. 6.2) with turbulent CO_2 flux showing net uptake during

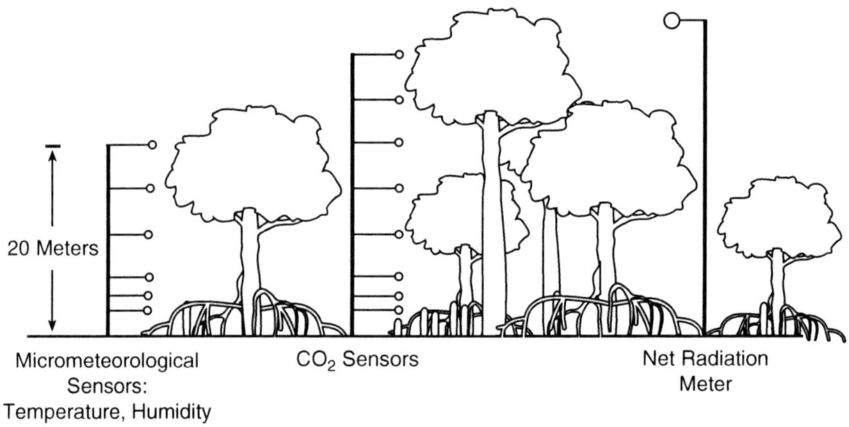

Fig. 6.1 Arrangement of the instrumentation and measurements needed to quantify the vertical gradient of CO_2 in a mangrove forest using the eddy covariance approach

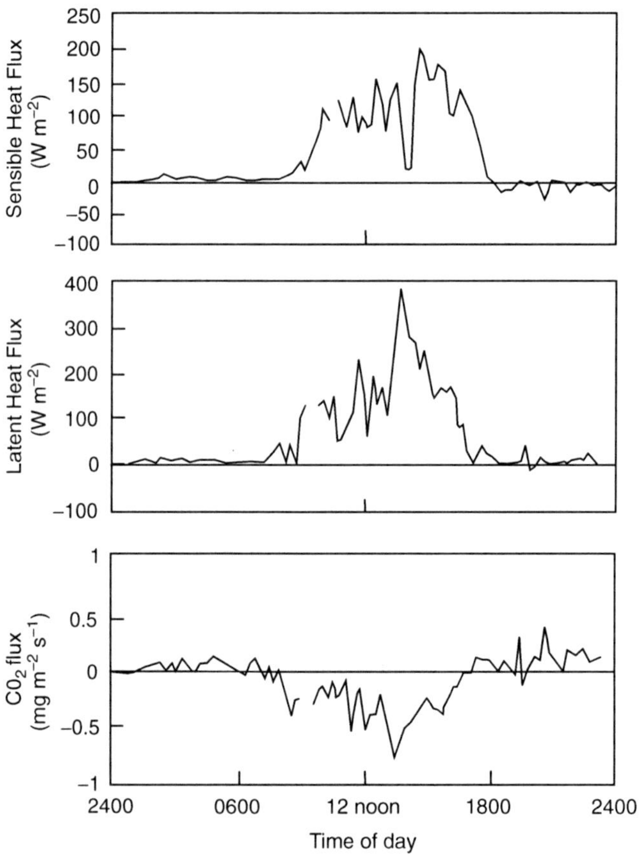

Fig. 6.2 Diurnal fluxes of sensible heat, latent heat, and CO_2 over the Phangnga mangrove forest of southern Thailand on 9 September 1998 (Modified from Monji et al. 2002a)

daytime (shown as a negative flux) and net flux out of the forest at night (a positive flux); this pattern is mirrored in the sensible and latent heat fluxes (Fig. 6.2). Heat energy mirrors gas flux because energy is used for the work of photosynthesis and respiration; latent heat is the energy transferred between the forest and the atmosphere by water evaporation or the condensation of water vapor, whereas sensible heat is the energy transferred between the forest and the atmosphere by conduction and movement by convection.

Soil respiration was an order of magnitude smaller than CO_2 flux from the canopy, with CO_2 flux not significantly different between wet and dry seasons in these Thai forests (Monji et al. 2002a, b). Net CO_2 flux averaged 0.11 mg m^{-2} s^{-1}. This equates to a net ecosystem production (excluding any tidal losses) of 78.8 mol C m^{-2} $year^{-1}$ and for the entire mangrove ecosystem at Phangnga (30,000 ha) of 23.6 Gmol C $year^{-1}$. A similar study conducted in the Sunderbans mangrove forest ecosystem bordering the northern Bay of Bengal (Ganguly et al. 2008) measured heat and gas fluxes over the course of a year. Ganguly et al. (2008) found a seasonal pulse of CO_2 flux, with lower rates of daytime flux during the wet monsoon (Fig. 6.3) which was attributed to cooler temperatures, a decline in salinity, and less solar radiation. Nighttime fluxes did not vary seasonally (Fig. 6.3) but the total CO_2 exchange was 121 mol C m^{-2} $year^{-1}$ for a total net ecosystem production for the Sunderbans (426,400 ha) of 515.9 Gmol C $year^{-1}$.

6.3.2 *The Mass Balance Approach*

The mass balance approach is a less recent and arguably more laborious (and expensive) method of estimating carbon balance and net ecosystem production. This approach is based on the simple premise that the flow of carbon in an ecosystem is in steady-state. A simple mass balance equation is the basis for the flux of

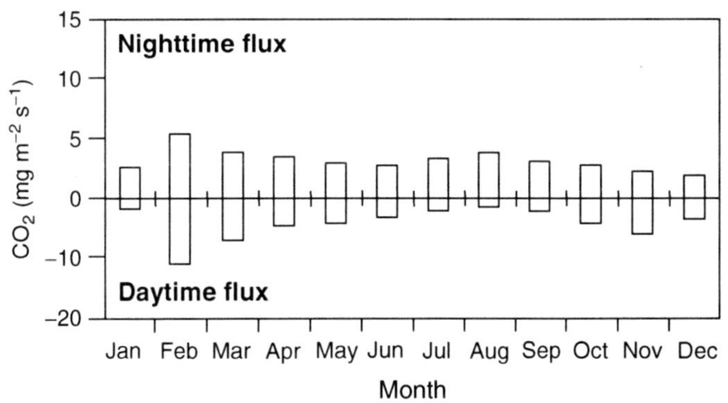

Fig. 6.3 Monthly variation of CO_2 flux during the day and at night over the Sunderbans mangrove forest. Vertical bars represent mean flux rates. (Modified from Ganguly et al. 2008)

carbon (or any other element), $C_i = F_i + \Sigma R_{I,}$, where C_i = concentration of element i in mass per unit volume per unit time; F_i = flux of element i in mass per unit area or volume per unit time; R_i = rates of each physical, chemical or biological process affecting element i in mass per unit volume or area per unit time. Simply, the equation represents the difference between what carbon comes in and what carbon goes out. In ecological terms, when carbon that comes in or is produced (total input) exceeds what carbon goes out or is consumed (total output), net ecosystem production is greater than zero and the system is accumulating carbon. If it roughly equals zero, the system is in steady-state and if negative, the ecosystem is losing more carbon that it is gaining. Whether or not ecosystems are in positive or negative balance has important consequences for ecosystem sustainability, as we will discuss in Section 6.7.

Mainly because so many separate measurements are involved, only six mangrove ecosystems have been studied sufficiently to enable preliminary mass balance estimates of carbon to be made: Rookery Bay in Florida, the Matang Mangrove Forest Reserve in Malaysia, Sawi Bay in Thailand, and Hinchinbrook Channel, Missionary Bay, and Darwin Harbour in northern Australia.

A few characteristics of carbon flow in these six ecosystems warrants analysis (Table 6.4): (1) these mangrove ecosystems are net autotrophic, with an average P_G/R ratio of 1.6, higher than the Gattuso et al. (1998) estimate of 1.4, and (2) gross primary production and net ecosystem production average 383 and 139 mol C m^{-2} $year^{-1}$, higher than the Gattuso estimates of 232 and 89 mol C m^{-2} $year^{-1}$, respectively. Despite the obvious drawbacks of such budgets, it is clear that:

- These ecosystems export organic carbon equivalent to 2-25% of mangrove net primary production.
- Canopy respiration equates to 58% of gross primary production and is probably higher as the data account only for respiration of leaves and do not include stem and root respiration.
- Mangrove production dominates carbon input, but inputs from human settlements and activities, and from riverine and oceanic contributions, can be substantial as exemplified in the Matang Mangrove Forest Reserve, Hinchinbrook Channel and Darwin Harbour ecosystems.
- Soil and water-column respiration losses pale in comparison to canopy respiration.
- There is proportionally little carbon burial within the forest floor, ≈1–4% of total organic carbon input to the forest.
- Net ecosystem production (NEP) is positive in all six ecosystems, but true NEP is probably lower because losses of CO_2 from the water to the atmosphere were not measured at any of these locations.

The differences between these estimates and those of Gattuso et al. (1998) are undoubtedly due to the dominance of Australasian ecosystems in the present calculations. These ecosystems are likely to be among the more productive mangroves in the world. Further, some values in Table 6.4 represent only a few measurements, so the results must be treated cautiously as there is large error in many of the numbers.

Table 6.4 Physical characteristics and mean rates of organic carbon inputs and outputs for the Rookery Bay (RB), Matang Mangrove Forest Reserve (MMFR), Sawi Bay (SB), Hinchinbrook Channel (HC), Missionary Bay (MB) and Darwin Harbour (DH) mangrove ecosystems. Units are mol C m−2 year−1, unless noted otherwise. Data from Twilley (1985b, 1988), Caffrey (2003), Clough et al. (1997b), Clough (1998, unpublished data), Gong and Ong (1990), Ayukai and Miller (1998b), Ayukai et al. (1998, 2000), Alongi (1998), Alongi et al. (1998, 1999, 2000c, 2001, 2004a), Tanaka and Choo (2000), Alongi and McKinnon (2005), Burford et al. (2008).

	RB	MMFR	SB	HC	MB	DH
Percentage of mangrove area to total ecosystem area	24%	67%	20%	36%	39%	18%
Tidal range (m)	0.55	2.0	1.3	2.4	2.3	7.8
Rainfall (m $year^{-1}$)	1.3	2.5	1.2	2.5	2.5	2.2
Sources						
Mangrove GPP	276.5	415.3	450.4	370.3	294.0	490.2
Algal GPP	68.5	69.2	16.1	16.6	16.4	49.9
Other inputs[a]	NA	23.2	0.1	15.8	NA	42.0
Total inputs	345.0	507.7	466.6	402.7	310.4	582.1
Sinks						
$Resp_{canopy}$	184.9	293.2	297.1	190.3	158.0	201.8
$Resp_{water}$	114.0	11.5	24.3	8.5	8.2	10.5
$Resp_{soil}$	16.4	36.1	26.7	9.4	10.7	53.6
Burial	NA	9.4	22.9	5.6	4.2	NA
Export and other outputs[b]	5.3	30.5	5.9	10.4	27.7	26.7
Total outputs	320.6	380.7	354.0	224.2	208.8	292.6
NEP[c]	24.4	127.0	112.6	178.5	101.6	289.5
%NEP/GPP	9%	31%	25%	48%	35%	59%
%NEP/NPP	27%	100%	73%	100%	75%	100%
P_G/R	1.09	1.42	1.34	1.86	1.75	2.03
% TOC input buried	NA	2%	4%	1%	1%	NA
% Export of mangrove NPP	6%	25%	2%	6%	20%	9%

[a]Net river and oceanic inflow, aquaculture wastes, sewage, seagrass production.
[b]Logging, net river and oceanic inflow, aquaculture wastes, sewage.
[c]Total Inputs minus Total Outputs.

Further, both Rookery Bay and Darwin Harbour represent outliers at both extremes of tidal range and are not typical of other mangrove estuaries, most of which reside in more moderate (1–3 m) tidal conditions.

What regulates organic carbon fluxes at the ecosystem level? Physiological constraints and physicochemical conditions control growth and production of individual populations and communities, but why does net ecosystem production, for example, vary so much among these six ecosystems? Characteristics unique to each ecosystem play some role. In the Matang Reserve, large tracts of the forest are sustainably clear-felled for fuel and charcoal, and there are less well-regulated aquaculture activities and organic pollution within the reserve. In Sawi Bay, most of the forests are <25 years old, as many stands have been restored after unregulated logging for pond aquaculture in the 1980s and for industrial developments in the 1970s.

A plot of the relationship between tidal range and net ecosystem production (Fig. 6.4) for these six ecosystems suggests that tides play an important role in regulating ecosystem production. This idea supports the tidal subsidy hypothesis originally exposed by the Odum's (Odum 1968, 2000; Odum et al. 1979, 1995) and by Nixon (1988). Odum et al. (1995) suggested that maximum power is achieved when biological 'pulses' are in synchrony with external pulses such as tides. A comparative analysis of lake and marine ecosystems led Nixon (1988) to argue that the additional mechanical energy from tides has important consequences for differences in functioning between freshwater and marine systems, including more intensive fisheries yield, stronger currents, and more vigorous vertical mixing. These physical factors are linked to tides, assisting in maximizing transport of wastes and toxic materials from the ecosystem, assisting in oxygenating soils that would otherwise be waterlogged, and maintaining an intermediate level of disturbance in breaking down biological, chemical, and physical gradients within estuaries. Such suppositions may serve to explain the positive relationship between tidal range and mangrove net ecosystem production.

A contrary hypothesis is that these values do not necessarily reflect net ecosystem production, but in fact represent losses of respired carbon not measured in these ecosystems. As we will discuss in Section 7.1, much carbon from soil respiration may be unaccounted for, lost via lateral transport or via groundwater flow, pathways that were not measured in these ecosystems. The large tidal range in Darwin Harbour, for example, may translate into greater lateral transport of respired carbon in the intersti-

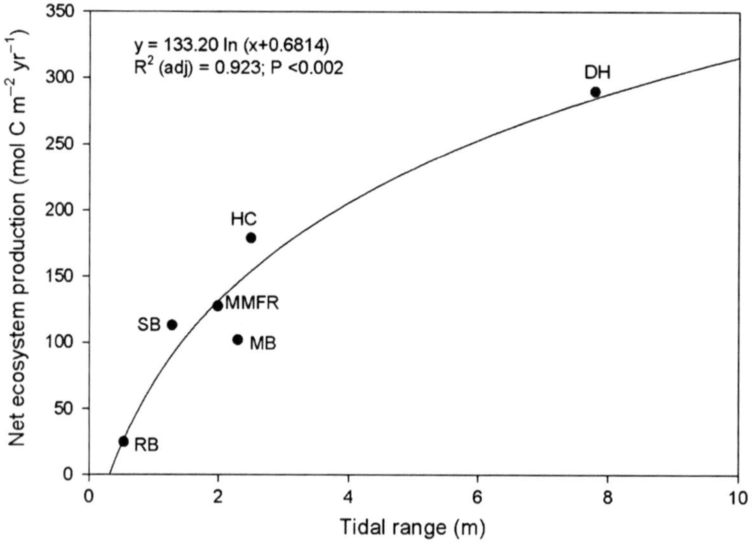

Fig. 6.4 The relationship between tidal range and net ecosystem production in various mangrove estuaries. (Data from Table 6.4). DH = Darwin Harbour, HC = Hinchinbrook Channel, MB = Missionary Bay, MMFR = Matang Mangrove Forest Reserve, SB = Sawi Bay, and RB = Rookery Bay

tial water from the forest floor; hence, what would appear to be greater NEP than in the ecosystems with smaller tides, may in fact be a greater proportion of carbon lost from the ecosystem. If true, this means that mangroves are contributing even greater amounts of dissolved inorganic carbon to the tropical coastal ocean (see Section 7.1).

6.4 Nitrogen Flow Through Mangrove Ecosystems: The Hinchinbrook Island Study

There is only one mangrove ecosystem for which a complete nitrogen budget exists: the mangroves of Missionary Bay at the northern end of Hinchinbrook Island in north Queensland, Australia (Fig. 6.5). The budget for this ecosystem was first presented in 1992 (Alongi et al. 1992), and except for some additional data and corrections, the initial conclusions remain valid. Nevertheless, it is instructive to update the budget and reiterate the conclusions here as they have greater relevance now as mangroves are becoming increasingly impacted by anthropogenic inputs.

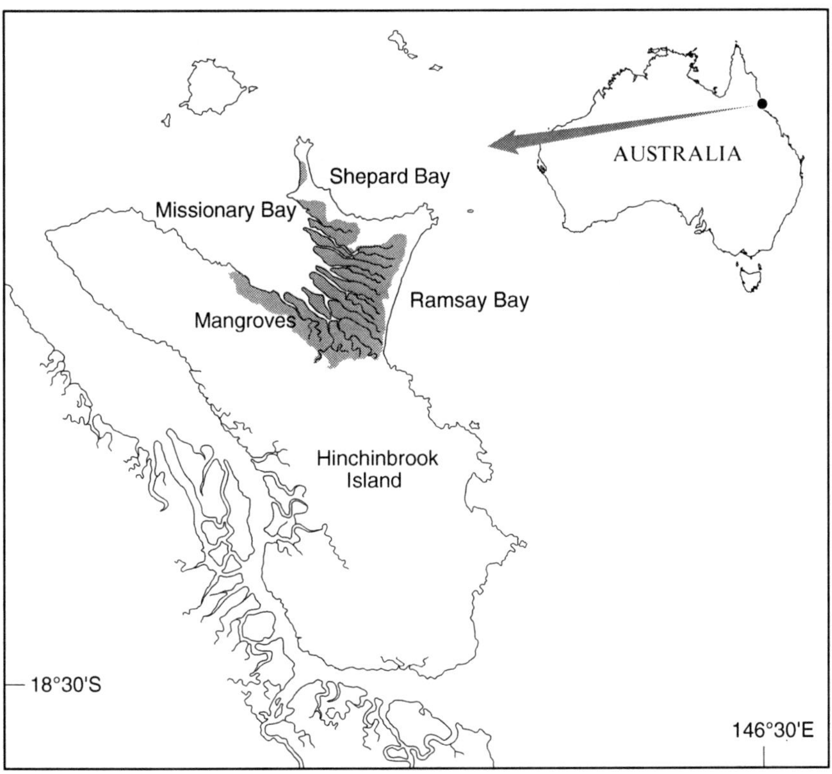

Fig. 6.5 Location of Hinchinbrook Island and the mangroves of Missionary Bay in relation to the Queensland coast

Hinchinbrook Island (18°20' S, 146°10' E) is a heavily forested and mountainous island that is a national park, situated adjacent to the Australian mainland about 100 km NW of the city of Townsville. The western boundary of the island is separated from the mainland by Hinchinbrook Channel, a narrow waterway 50 km long and lined with extensive, deltaic mangrove forests. At the northern boundary is located Missionary Bay (64 km^2), consisting of a series of finger-like projections of mangrove forests separated by tidal creeks. The entire mangrove area is 42.5 km^2, composed mostly of mixed *Rhizophora* spp. Total water volume exchanged over an average tidal cycle is approximately 1.5×10^7 m^3; tides are semi-diurnal with an average range of 2.3 m. Although the Missionary Bay mangroves receive an average annual rainfall of 2,500 mm, there is no significant groundwater input. It is a fully marine system. Even during the summer wet season, salinity rarely dips below 33.

The main pathways for nitrogen to enter the system are: (1) nitrogen fixation by microbial consortia residing on surface soils, logs, other pieces of fallen timber, and live tree stems and above-ground roots, (2) tidal exchange, and (3) precipitation (Table 6.5). The estimate for groundwater represents the amount of rainwater falling within the mountainous barrier bordering the mangroves and assumes that all of this water runs down to the mangroves. Tides bring in mainly DON with lesser amounts of ammonium, nitrate and nitrite, but no particulate nitrogen.

Net tidal exchange represents losses due mainly to the export of litter (Table 6.5) which constitutes the main loss of nitrogen from the system, followed by denitrification and to a lesser extent, sedimentation in the forest and on creek banks. The outwelling of particulate and dissolved nitrogen is clearly the dominant process. Other possible sources and sinks remain unquantified, such as volatilization of ammonium, and migration of fish, prawns, birds, and bats, but these losses are probably small and unlikely to significantly alter the net balance of nitrogen.

Net tidal exchange is 97 Kmol N $year^{-1}$. Considering the extrapolations made and the systematic and relative errors involved with so many different individual measurements over time, the budget is in overall balance. Considering that the net ecosystem production is estimated at 4,318,000 Kmol C $year^{-1}$ (Table 6.4), it is clear that nitrogen is greatly conserved within this ecosystem.

A number of mechanisms operate to conserve nitrogen. First, there is a very high rate of N cycling within the soil, to the extent that the bulk of dissolved nitrogen is taken up by the trees; as noted in Chapter 5, comparatively little (about 5% of total N input to the soil) is lost via denitrification and by efflux during flood tides. The microbe–soil–root complex thus rapidly recycles nitrogen (and other nutrients) via mortality, decomposition, uptake, and growth of organisms, thus serving as a retention mechanism. Second, crabs assist in minimizing the loss of litter and maximizing N gain by their activities in the soil. Third, tree stems, roots, logs, and other mangrove timber on the forest floor provide space to maximize colonization of nitrogen-fixers and thus the rate of atmospheric input. Finally, the C:N ratio of dissolved and particulate material leaving the forest is higher than Redfield and high in concentrations of refractory compounds such as humic and fulvic acids, and polyphenolics, indicative of an advanced state of decomposition. In short, little that is biologically labile leaves the forest and its waterways.

Table 6.5 Nitrogen budget (kmol N $year^{-1}$) for the mangrove ecosystem in Missionary Bay, Australia. The budget was constructed using data, including habitat areas, from papers cited on page 279 in Alongi et al. (1992)

Process	Input	Output	Net exchange
Precipitation			
$NO_2 + NO_3$	0.7		
NH_4^+	0.5		
DON	1.3		
Particulate N	0.1		
			2.6
Groundwater			
	2.4		
			2.4
Nitrogen fixation			
Saltpan	466.2		
Soil surface	479.7		
Prop roots	1,192.7		
Logs/timber	930.7		
Stem	376.7		
			3,446
Tidal exchange			
$NO_2 + NO_3$	437.5	525.0	−87.5
NH_4^+	928.0	696.8	231.2
DON	12,684.3	8,821.4	3,862.9
Particulate N		6,360.8	−6,360.8
Denitrification			
		658.4	−658.4
Sedimentation			
		342.5	−342.5
Total	17,501	17,405	95.9

As no other mangrove ecosystem has been explored in sufficient detail to enable construction of a nitrogen budget, comparisons with other mangroves are limited to specific processes. For instance, in the Fukido mangroves on Okinawa (Kurosawa et al. 2003), export of nitrogen was measured as 0.17 mmol N m^{-2} day^{-1} roughly equivalent to accumulation of nitrogen in the soil (0.2 mmol N m^{-2} day^{-1}). In this ecosystem, both export and accumulation each equaled only about 5% of gross primary production. In the Potengi mangrove forests of northern Brazil (Silva et al. 2007), export of N was estimated at 0.3 mmol N m^{-2} day^{-1}, and like the Fukido mangroves, this export constituted only a small percentage of mangrove production.

One process that has only recently been measured from mangrove ecosystems is the loss of ammonia gas. This process was assumed to be relatively insignificant in Missionary Bay, but a thorough study of ammonia exchange in the Sunderbans suggests otherwise (Biswas et al. 2005). Emission of ammonia gas from the forests to the atmosphere averages 1,790 kgN km^{-2} $year^{-1}$ but wet and dry inputs average 2,350.5 kg N km^{-2} $year^{-1}$ and there is net import of ammonia to the waterways on the order of 775.7 kg N km^{-2} $year^{-1}$. If one applies these rates to Missionary Bay, there would be a total input of 7,910.7 kmol N year^{-1} and total output of 5,434 kmol year^{-1} for a total net ammonia exchange of nearly 2,500 kmol N year^{-1}, which would be the second largest flux after DON exchange (Table 6.5).

As indicated by Alongi (1998), there are a few insights to be learned from comparing the nitrogen budget of Missionary Bay and the nitrogen budget for the Great Sippewissett salt marsh located on the western shore of Cape Cod, Massachusetts. We can how update this information and add the nitrogen budget for another salt marsh, the Sapelo Island ecosystem, located on the Georgia coast (Table 6.6). There are a number of similarities between the mangroves and salt marshes:

- Tides dominate physical control of nutrient exchange.
- There is an overall balance between inputs and outputs, considering magnitude of errors.
- Tidal import and nitrogen fixation are the major inputs.
- Tidal outwelling is the major output.

But there are a number of striking differences:

- Nitrogen derived from freshwater is proportionally smaller in Missionary Bay.
- Tidal export as a percentage of total output is greater in Missionary Bay.
- Denitrification is a greater proportion of total output in the Great Sippewissett (12%) and Sapelo Island (25%) marshes than in the mangroves (4%).
- Rates of nitrogen fixation exceed denitrificatiion in Missionary Bay, but the reverse is true in the salt marshes.
- Sedimentation rates are confounded between ecosystems, with Missionary Bay and Sapelo Island having proportionally small (1–2% of total output) burial rates compared with Great Sippewissett marsh (12%).
- There is little particulate input into Missionary Bay, but there is significant import of particulate nitrogen into both salt marshes.

Table 6.6 Comparison of nitrogen budgets (kg N year^{-1}) of the Australian mangrove and American salt marsh ecosystems (Data from Table 6.5, Valiela and Teal 1979; Whitney et al. 1981; Thomas and Christian 2001). NA = not available

	Missionary Bay, Queensland, Australia	Great Sippewissett, Massachusetts, USA	Sapelo Island, Georgia, USA
Inputs			
Precipitation	36	271	3,480
Groundwater	34	6,435	464
N_2 fixation	48,244	1,642	212,052
Tidal import	196,697	21,833	740,560
Other	0	20	NA
Total inputs	245,011	30,201	956,556
Outputs			
Tidal export	229,656	26,316	762,352
Denitrification	9,218	4,349	261,864
Sedimentation	4,795	4,150	15,086
Other	0	30	NA
Total outputs	243,669	34,845	1,039,302
Net exchange	+1,342	–4,644	82,746

It is unlikely, however, that such differences and similarities in nitrogen flow will hold true in any future comparisons between mangroves and salt marshes. Other mangroves and salt marshes are undoubtedly different, and it is exceedingly likely that such comparisons will find differences among mangrove ecosystems as great as those between mangroves and salt marshes. We do not have sufficient understanding of the inherent functional characteristics of each ecosystem type, as more ecosystem-level budgets are urgently needed to redress this discrepancy.

6.5 Mineral Cycling

With the exception of some data detailing element concentrations in soil and tree components, little information is available on cycling of essential elements such as iron, potassium, sodium, magnesium, calcium, zinc, copper, manganese, and vanadium, in mangrove ecosystems. Work in Chinese mangroves by Lin Peng and his colleagues (Li 1997; Lin 1999) have established turnover times of various elements via litterfall. Litter represents only a small part of the cycling of matter in a mangrove forest, but these extensive studies offer a glimpse of variations in turnover times among species and elements (Fig. 6.6). The turnover times of potassium, calcium, and sodium are shorter in *Kandelia candel* forests than through the litterfall of *R. stylosa* and *B. sexangula* forests. Importantly, these data suggest that turnover of these elements

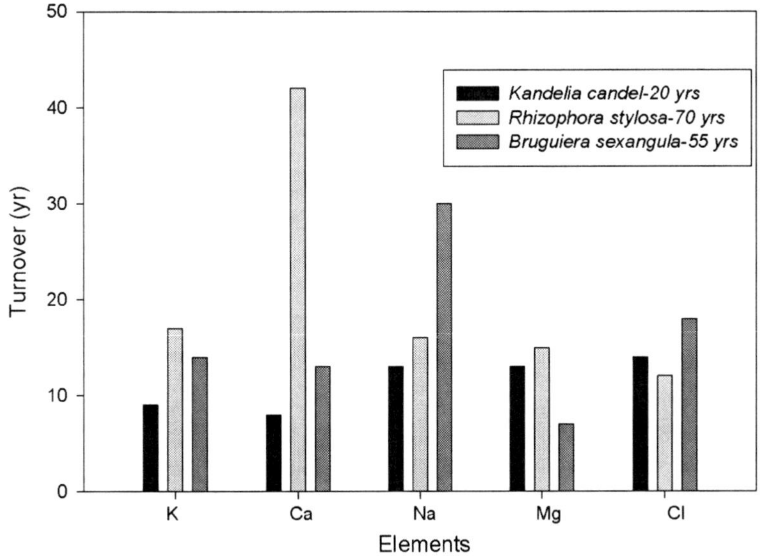

Fig. 6.6 Turnover time of potassium, calcium, sodium, magnesium, and chlorine as throughput via litterfall in three mangrove forests of different species and age in southern China (Data from Lin 1999)

Table 6.7 Turnover times (years) of soil elements in four mangrove forests of different age in southern Thailand (Modified from Alongi et al. 2004b)

	25 year-old *R. apiculata*	5 year-old *R. apiculata*	3 year-old *R. apiculata*	3 year-old *C. decandra*
Forest production (t C ha^{-1} $year^{-1}$)→	52.8	16.9	37.0	22.9
Element				
N	4	13	12	16
P	6	26	13	22
S	113	642	322	169
Fe	916	21,272	18,264	23,098
Na	6	43	36	50
Mg	9	85	81	65
K	21	127	64	114
Ca	2	40	8	17
Zn	162	733	505	453
Cu	610	470	147	275
Mn	3	120	14	139
Mo	28	320	430	200

is faster in mangroves, on average, than in many terrestrial forest types (Barnes et al. 1998; Aber and Melillo 2001; Kimmins 2004).

The turnover time of an element most likely relates more closely to the rate of primary production than to forest age. For instance, in mangrove forests aged from 3 to 25 years in southern Thailand, turnover of soil nitrogen, phosphorus, sulfur, iron, sodium, magnesium, manganese, molybdenum, potassium, calcium, copper, and zinc is shorter in the oldest forest, which is also the most productive (Table 6.7). Simply, the greater the rate of carbon fixation, the greater the demand for essential elements. The turnover times in the Chinese mangroves are roughly on the same scale as found in other tropical wetlands and forests (Golley et al. 1975; Dykyjova and Ulehlova 1998; Fassbender 1998). However, complete ecosystem-level budgets of many essential elements are urgently needed to offer insights into element cycling in mangrove forests.

6.6 Systems Analysis: Understanding Links Among Various Functions of an Ecosystem

Systems ecology as nurtured by Howard Odum (1983) has led to the development of conceptual and stimulation models into the behavior of the functioning of ecosystems. Such models are used to highlight gaps in our knowledge and to predict the impact of pollutants and other human disturbances on ecosystems, but they are being used increasingly for economic purposes in the relatively new field of ecological economics. In this section, models depicting various functions of mangrove ecosystems are assessed in order to understand the use of energetics information in trying to quantify sustainability.

6.6.1 Network Models

The first model of a mangrove ecosystem was conceptual, fostered by the need to clarify new views of trophic interactions and detritus flow (Heald 1969; Odum and Heald 1975). The first true computer simulation model of a mangrove ecosystem was constructed by Ariel Lugo and his colleagues (Lugo et al. 1976). In fact, the original model continues to be improved upon (Pandey and Khanna 1998). In this model, the main pathway of energy flow is the export of detritus from the forest floor to the estuary, with some material being lost through grazing and decomposition, and intercompartmental transfers and interactions modelled as either linear or non-linear functions, depending on the process and assuming steady-state. The model and subsequent sensitivity analysis (Pandey and Khanna 1998) indicates that all variables (detritus, nutrients, mangrove biomass, etc.) are highly sensitive to tidal action, with the most pronounced effect being on the flow of detritus. The strongest influence on mangrove biomass is terrestrial run-off and, secondarily, solar radiation.

An attempt to incorporate modelling of litterfall dynamics into an energy flux budget was first attempted by Wafar et al. (1997) for the mangroves of the Mandovi and Zuari estuaries in west India. They found that mangrove production is important mainly in carbon flow with respect to microbial decomposition, rather than in food chain dynamics directly. Development of computer packages such as NETWORK (Kay et al. 1989) and ECOPATH (Ulanowicz and Kay 1991) has led to use of such models for more sophisticated analysis of flows in mangrove ecosystems (Manickchand-Heileman et al. 1998; Lin et al. 1999; Ray et al. 2000; Wolff et al. 2000; Vega-Cendejas and Arreguin-Sánchez 2001; Ortiz and Wolff 2004; Cruz-Escalona et al. 2007; Ray 2008).

The main premise of these models is to solve a steady-state, mass balance-type equation and to calculate the following network characteristics from the results: the mean length of energy flow pathways, the percentage of recycling (Finn's cycling index, Finn 1976), the extent of connectivity, detritivory, and herbivory among trophic groups, and total system throughout (sum of all energy flows). This is accomplished by balancing the linear equation, $B_i P_i (B_i)^{-1} E_{Ei} - {}^{n}\Sigma_{j=1} B_j Q_j (B_j)^{-1} D_{Ci} - E_{Xi} = 0$, where B_i = biomass of group i, $P_i B_i^{-1}$ = production/biomass ratio of i, E_{Ei} = proportion of production shunted into predation, B_j = biomass of predator j, $Q_j B_j^{-1}$ = consumption/biomass ratio of predator j, D_{Ci} = fraction of prey i in diet of predator j, E_{Xi} = export of group i.

Use of early versions of ECOPATH resulted in a number of energy flow and trophic transfer models for mangrove-dominated ecosystems. Analysis of Terminos Lagoon in Mexico (Manickchand-Heileman et al. 1998) shows that trophic transfer efficiency is low (7%) but that there is a high degree of recycling (Finn index = 7.0) and a long average length of food chain paths (10). These results indicate that most energy flow is vested in lower trophic levels. This is despite the fact that early versions of the software did not include bacterial utilization of detritus or possible burial of organic material in sediments; workers at that time were unaware of the preference for benthic algae over detritus by consumers so the model favors the dominance of detrital pathways within the system. In contrast, a balanced trophic model of another lagoon in southwestern Taiwan (Lin et al. 1999) found that high planktonic primary production is the main energy driver of food chains dominated by herbivorous zooplankton. Half of the carbon fixed by phytoplankton is not immediately used by higher trophic levels but shunted into the detritus pool, most of which is directly consumed by fish and other large consumers and eventually exported to intensive fisheries. These fluxes reflect short trophic pathways (mean path length = 3.38) with high trophic transfer efficiencies among the upper trophic groups.

Network analysis of two other coastal lagoons in Mexico (Vega-Cendejas and Arreguin-Sánchez 2001; Cruz-Escalona et al. 2007) indicates intense fisheries productivity, similar to that in Terminos Lagoon, but in one system (Celestun Lagoon) most primary production is exported, as only 4% is grazed and 7% is shunted into the detritus pool. In the other lagoon (Laguna Alvarado) as in Taiwan, net primary production is the main source of energy, with consumption, respiration, and detrital fluxes accounting for 47%, 37%, and 16% of total system throughput, respectively. Clearly, all of these models show that lagoonal-mangrove systems are highly productive, but also point to autotrophs other than mangroves as playing a major trophic role.

In more open, riverine estuaries, modeling of trophic flows supports empirical evidence that mangroves play the major role in energy flow (Wolff et al. 2000). Within the Amazon system, the Caeté tributary possesses one of the world's largest expanses of mangroves, but is subject to intense logging and harvesting of crabs. Modelling of this system has found that shrimp and fish are of relatively low energetic importance (compared with the lower trophic groups) along with mangrove epibenthos, especially crabs such as *Ucides cordatus*, which are heavily harvested. Mean trophic transfer efficiency (10%) and the high gross efficiency of the fishery (catch/net primary production = 9%) is explained by high rates of mangrove logging and crab harvesting. Bacteria contribute 34%, mangroves 19%, fiddler crabs 13%, algae 10%, mangrove crabs 10%, and the remaining trophic groups 14%, of total energy flow. Of greater significance is that the modelling suggests overexploitation of crabs; use of a realistic P:B ratio results in the model not balancing, that is, that more biomass is being harvested than produced. This scenario is a good demonstration of how models can be used to determine the level of sustainable harvesting of resources.

Such models are also of use in determining ecosystem-level differences between pristine and reclaimed mangrove forests. A good example is the work of Santanu Ray and his colleagues (Ray et al. 2000; Ray 2008) in the Sunderbans in India. In an initial network analysis of the benthic food web within the impacted mangroves, Ray et al. (2000) found that herbivorous and detritivorous pathways are equally important, with human impacts resulting in an increase in the relative importance of algal primary production, herbivorous zooplankton, and meiofauna. Recycling of material is low, but path redundancy is high, suggesting that the benthic communities residing in the impacted mangroves are probably highly resilient to further stress. Further modelling reveals that benthic communities within the virgin forests are typically more detritus-based, being more dependent on litter. The pristine forests have more throughout (539,040 kcal m^{-2} $year^{-1}$) than the reclaimed forests (136,570 kcal m^{-2} $year^{-1}$), indicating more energy is passing through a given area per unit time, but a greater percentage of energy is being lost via respiration in the disturbed forests (Fig. 6.7). Some other clear differences between ecosystems were detected in the network analysis (Fig. 6.7). Relative ascendency, percentage imports and exports, and the Finn cycling index, are all greater in the pristine than in the reclaimed mangroves. This means that the food webs in the reclaimed forests are less organized, but possess communities more capable of replacing each other within a given niche. Also, proportionally more energy is imported and exported from the healthy ecosystem with a greater percentage vested in recycling pathways. It appears that, even with a number of limitations and unsubstantiated assumptions of linearity and steady-state, these models are useful tools for analyzing salient characteristics of how mangroves function as ecosystems.

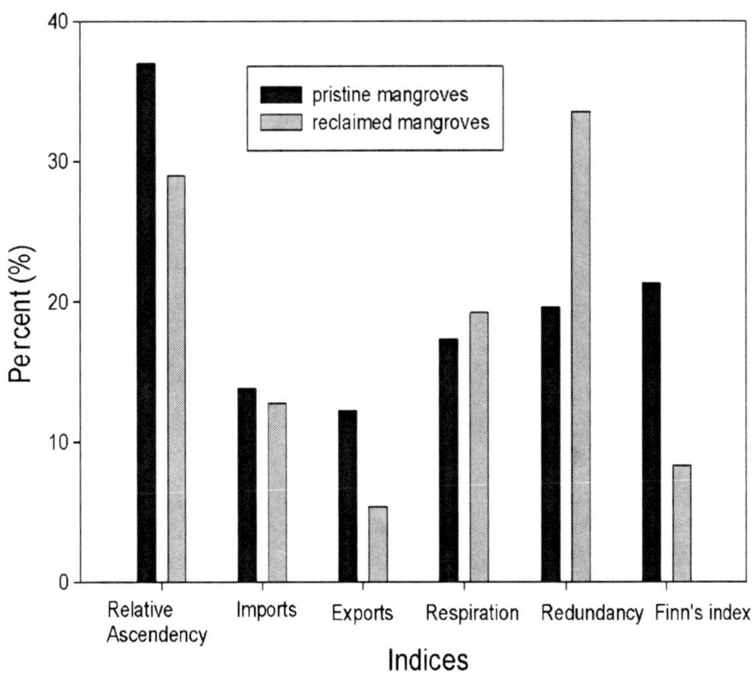

Fig. 6.7 Comparison of some key ecosystem characteristics between pristine and reclaimed mangrove forests in the Sunderbans, India, calculated using network analysis (Data from Ray 2008)

6.6.2 *Ecohydrology: Linking Physics and Ecology for Management Applications*

Recognizing the links between physical and biological characteristics within ecosystems, and the urgent need to apply ecosystem ecology to management issues, an international meeting sponsored by the International Hydrological Programme of UNESCO developed a rationale and conceptual framework for a new type of model in which hydrological processes are linked to ecological submodels (Zalewski et al. 1997). In the model, a physical submodel divides an estuary into a series of interlinking cells from the mouth to the upper tidal limit of the waterway (Fig. 6.8). The cell at the upper tidal limit receives Q_f, river discharge, sediment (Q_s), detritus, freshwater plankton, and nutrients. The cell at the estuary mouth receives ocean inputs of water, detritus, sediment, nutrients, plankton, and fish. There is a downstream flux resulting from the riverine input through the series of cells and a bidirectional flux from cell to cell depicting tidal mixing, resulting in a turbidity maximum and a limit of oceanic intrusion. There is also either a rate of import or export from every cell from mangroves. These processes are then fit to empirical data and simulated as a series of subroutines using steady-state equations of physical flows and predator–prey interactions.

The first use of such a model for mangroves was for Darwin Harbour, a large macrotidal estuary sheltering a small city in the Northern Territory, Australia

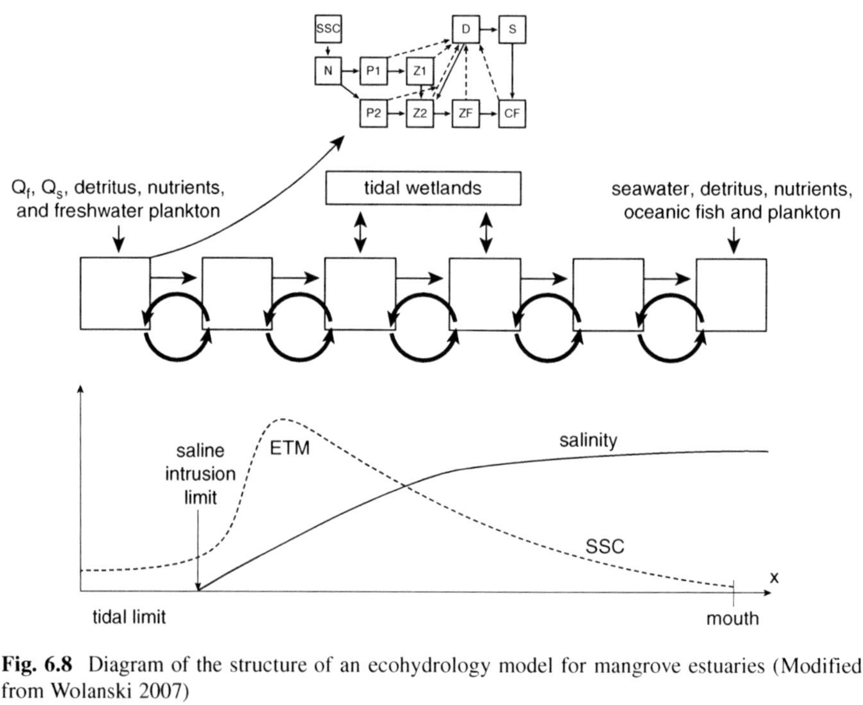

Fig. 6.8 Diagram of the structure of an ecohydrology model for mangrove estuaries (Modified from Wolanski 2007)

(Wolanski et al. 2006). This estuary is pristine but there are plans to expand the harbor and human encroachment into the estuary. The model was used to test three scenarios: (1) the impact of doubling suspended sediment concentrations through increased land clearing; (2) the impact of removing all mangroves; and (3) the impact of doubling dissolved nutrient concentrations in the upper and middle reaches of the estuary, but with no change in suspended sediment concentrations. Model outputs show that:

- In the first scenario, dissolved nutrient concentrations would quadruple in the upper half of the harbor, picoplankton abundance would increase by 50% in the middle region, total chlorophyll levels would quadruple in the middle reaches, numbers of copepod nauplii would be little affected, but adult copepods would triple in number in the middle region.
- In the second scenario, carnivorous and detritivorous fish numbers would decrease by 70% and 50%, respectively, in the upper and middle reaches of the estuary.
- In the third scenario, picoplankton numbers would increase by 50% in the upper and middle reaches, total chlorophyll concentrations would quadruple in the middle reaches, and copepod nauplii would increase by 30% in the upper and middle reaches.

The model thus suggests that ecosystem structure and function would be greatly impacted by large-scale changes in land-use.

A similar model was used to determine the fate of organic carbon in a Tanzanian mangrove ecosystem (Machiwa and Hallberg 2002). In this study, the model was composed of three submodels representing dissolved, particulate, and litter organic carbon, as well as tidal regime parameters and the size of macrofauna and microfauna communities. The model confirms empirical observations that DOC export is a major feature of this partially-impacted system, accounting for nearly 80% of organic carbon export. Forty percent of DOC is estimated to be utilized by microorganisms in order for the model to balance. The model confirms an initial premise that the extent of mangrove cover and hydrodynamic properties are strong determinants in the extent of organic carbon export from the ecosystem.

These models must be interpreted with caution, given the simplicity of the models compared with reality and the fact, in both cases, that the empirical data is inadequate for detailed calibration. Nevertheless, they are good starting points for applying ecosystems data to practical management problems.

6.7 Ecological Economics and Sustainability of Mangroves

The universal laws and concepts of ecological energetics (see Chapter 1) are analagous to some of the key principles of economics (Odum 1973; Smil 2008). The concept of ecology-economic commonality is indirectly intertwined with the fact that the energetics of ecosystems (e.g., productivity) is an important variable in humankind's needs and activities, and in the global economy. A simple illustration of the linkage is watts = joules second^{-1}, where power (W) is simply the rate of flow of energy (J s^{-1}). This follows on from the laws of thermodynamics which also incorporate the idea that the maximum amount of work that can be derived from a reaction is related to changes in temperature, the heat content of a system, and entropy (the degree of 'disorder' in a closed system). Energy, whether identified as power or the ability to do work, is a key driver in economic production (Smil 2008). Although the field of ecological economics is concerned with energy and its usage, it inevitably focuses on the issue of sustainability.

6.7.1 Models of Resource Economics

Mangrove ecosystems provide a large number of goods and services that are utilized by humans, both commercially and on a subsistence basis (Moberg and Rönnbäck 2003). Although most resource-use models analyze the trade-off in the cost to benefit ratio of the exploitation for wood and fisheries (Barbier and Strand 1998; Grasso 1998; Janssen and Padilla 1999; Larsson and Padilla 1999; Nickerson 1999; Rönnback 1999), a few models have integrated ecological and socioeconomic limits of human resource use (Twilley et al. 1998; Ortiz and Wolff 2004).

One of the first studies to investigate the impact of mangrove loss on sustainability of fisheries was conducted for Campeche, Mexico (Barbier and Strand 1998). Using a traditional catch-effort fisheries model with equations representing changes in fishing effort in response to profit margin and changes in mangrove area, Barbier and Strand (1998) found that even a small change in mangrove area greatly impacts the shrimp harvest (Fig. 6.9). The value of the mangrove habitat in supporting the shrimp fishery is thus affected by the level of exploitation. In the long run, the economic value of the fishery will be lower if it becomes heavily depleted.

A more sophisticated modeling effort was employed by Grasso (1998) for the similar purpose of determining the trade-offs involved in maximizing wood versus fish harvesting in Brazil. Manipulating the number of workers in each industry in the models, the simulation runs reveal that the most important variable is forest growth rate, with the balance in the number of workers in each industry also dependent on the relationship between price and stock size of the resources. For instance, declining forest size results in an increase in the price of wood, but the model predicts that the number of workers will decline, shifting over to fisheries-related work. However, there is a negative feedback loop in that fishery production is directly dependent on forest area.

A similar scenario has developed in Luzon in the Philippines in which mangrove area has declined because of multiple uses that are closely interconnected (Nickerson 1999). Running a population dynamics model with a cost-benefit analysis, Nickerson (1999) found that the conflicting needs of various users (fisherman, aquaculturists, foresters, traditional users, etc.) can best we met in the long term if the mangroves are left alone (Fig. 6.10). There is little difference in the

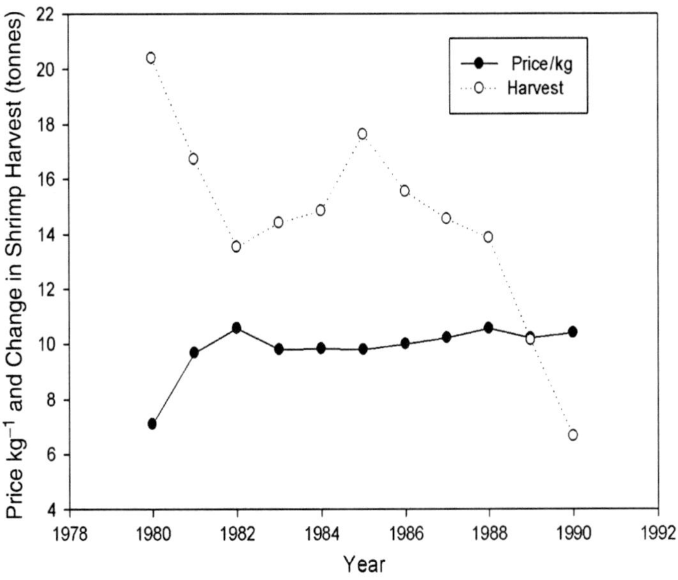

Fig. 6.9 Impact of a small decline in mangrove area (0.20%) on the price and level of harvesting of shrimp off the Campeche coast, Mexico (Data from Barbier and Strand 1998)

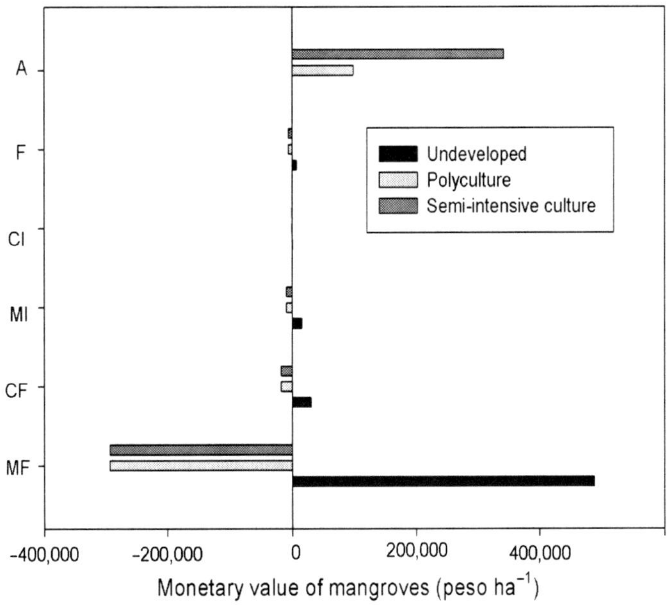

Fig. 6.10 Impact of (1) leaving mangroves undeveloped, (2) developed for polyculture of milkfish, and (3) developed solely for shrimp aquaculture, on the monetary value of mangroves for municipal fisherman (MF), commercial fisherman (CF), municipal families using invertebrates (MI), commercial users of invertebrates (CI), foresters (F) and aquaculturists (A) in Luzon, Philippines (Data from Nickerson 1999)

value of mangroves if used either for development of milkfish polyculture or for shrimp pond aquaculture, as both scenarios greatly lessen the value of mangroves compared with leaving the mangroves undeveloped. Unfortunately, this model and another (Janssen and Padilla 1999) do not reconcile the tendency to favor short-term profit over long-term ecological and economic gain, as these areas continue to lose mangroves to human development. Mangroves continue to be undervalued economically (Rönnbäck 1999).

Models that have integrated ecological and economic constraints on unsustainable versus sustainable practices suggest more realistically that social and cultural policies must be considered (and often altered) to reconcile competing needs. For the Guayas estuary in Ecuador, Twilley et al. (1998) developed simulation models of land use to demonstrate the impact of the rate and intensity of human use on the environmental quality of the estuary. The model runs show that a 90% decline in mangrove area would result in a five fold increase in nitrogen concentrations; construction of a dam upriver, however, would result in a 60-fold increase in nutrient concentrations and a shift in intertidal and upland zones due to the decline in river discharge. A more recent model for the Caeté estuary in Brazil (Ortiz and Wolff 2004) indicates that exploitation of mangroves for wood and fisheries is currently unsustainable; a rotation harvest system is recommended for mangrove harvesting. While estimation of maximum sustainable yield of mangrove wood and other resources is urgently needed, no model can alter the fact that humans conserve what is most monetarily valuable in the short-term, especially in areas where income is limited.

6.7.2 Using Ecosystems Data to Quantify Sustainability

So how do we quantify sustainability in a meaningful way? Can the type of information and energetics approach discussed throughout this book be useful for this purpose? In order to develop strategies for sustainable management, it is important to quantify limits to sustainable harvesting, including the level of ecosystem support required to maintain production. Here we examine in more detail the managed *Rhizophora apiculata* forests in Thailand and Malaysia discussed earlier (Section 6.3.2) to show how simple models of carbon mass balance may be used as a management tool to provide preliminary estimates of sustainable wood harvesting and ecosystem support.

Strategies used to achieve sustainable timber production in tree plantations range from the application of traditional silviculture techniques to whole-forest-scale dynamic models of maximum sustainable harvesting (Fujimori 2001; Nyland 2001). Strategies applicable at the ecosystem-level have recently gained favor with silviculturalists and plantation managers in recognition of the fact that tree growth is ultimately linked to, and supported by, a host of factors operating beyond the physical limits of a particular forest. This landscape approach is based on the idea that society values forests beyond single factors such as timber production, also valuing forests for their diversity, health, and aesthetics (Rowe 1994). Strategies based on this concept have proven successful in conserving and sustaining terrestrial forests and plantations (Raison et al. 2001).

The increase in global demand for tropical wood (Brown et al. 1997) comes at a time of declining wood yield due to environmental degradation, unsustainable harvesting practices, disease outbreaks, soil erosion, poor planning, and shortage of fresh water. It is possible that an increase in current production capacity of mangrove plantations will be needed to increase the global supply of hardwood. Mangroves are of high ecological and economic value, and their future may have to partly rely on an ecological economic approach rather than solely on conservation, especially in developing nations. A number of mangrove species are good candidates for cultivation because of their high rates of wood production. A number of plantations currently harvest mangroves, but few adhere to sustainable management practices (Saenger 2002).

A major impediment to developing strategies that can lead to sustainable timber harvesting is the lack of integrated ecological information, such as combining knowledge of factors limiting tree growth and resource use efficiency with management plans on stand rotation. Most useful techniques currently in use are based on traditional trial-and-error methods rather than on empirical scientific information.

As discussed earlier, several ecological and economic models currently exist to determine the interactive dynamics, growth, and yield of mangrove forests (Devoe and Cole 1998; Fromard et al. 1998; Berger and Hildenbrandt 2000) and to determine trade-offs between mangrove resource use and the economics of conservation (Ruitenbeek 1994; Grasso 1998; Janssen and Padilla 1999; Rönnbäck 1999; Huitric et al. 2002). Also, there have been at least two attempts to estimate the level of ecosystem support, or ecological footprint (*in sensu* Chambers et al. 2000), of mangrove ecosystems required to sustain coastal aquaculture (Robertson and Phillips 1995; Larsson and Padilla 1999). Despite this information, there are no quantifiable estimates of limits to sustainable production of mangroves or of the ecological footprint required for sustainable harvesting of mangroves. Mathematical and computer simulation models exist for estimating maximum sustainable yield (MSY) of temperate forests (Fujimori 2001; Nyland 2001), but the ecological information required to develop useful algorithms incorporating complex linkages and feedbacks for mangrove forests is insufficient.

6.7.2.1 Ecosystem-Scale Consequences of Forest Decline

While ground-truthing and GIS technology make it feasible to document habitat losses, the ability to estimate the timeframe of possible ecosystem collapse is difficult. Here we discuss how mass balance models of carbon flux in mangrove ecosystems could be used to: (1) develop estimates of sustainable wood production, (2) estimate the level of ecosystem support required to maintain yield, and (3) develop time frames for managers to avoid possible ecosystem collapse in the face of unsustainable harvesting. Data from Sawi Bay in southern Thailand and the Matang Mangrove Forest Reserve in northwestern peninsular Malaysia are used to illustrate.

The concept is simple, based on the laws of thermodynamics as applied in ecosystems ecology. If net ecosystem production is zero, the ecosystem is considered at the limit of sustainability. That is, the amount of organic matter produced or imported is equal to the amount of organic matter consumed or exported. If less than zero, the ecosystem is energetically unsustainable, at least over the long term. This is because an ecosystem that over time loses more organic matter (and energy) that it produces or gains, cannot exist for long; it will eventually disappear to be replaced by another type of ecosystem.

The term sustainability is vaguely defined in the ecological literature (Phillis and Andriantiatsaholiniana 2001), but is defined here in its simplest sense: to maintain a level of exploitation or production by restricting the quantity harvested (or extracted) to avoid long-term depletion.

A mass balance of carbon was constructed (Table 6.4) for Sawi Bay, a mangrove-fringed embayment in Chumphon province of southern Thailand. The bay is 165 km^2 in area, shallow, and open to the Gulf of Thailand. The watershed and mangrove ecosystem (mostly *Rhizophora apiculata*) have seen major changes since the 1970s – increases in mussel culture, fisheries, agricultural and industrial activities, and sewage, but particularly growth of shrimp ponds (Ratanasermpong et al. 2000). The increase in both aquaculture and agriculture has resulted in severe losses of mangroves. The Royal Forestry Department now has a replanting program in place, but clearing of mangroves and adjacent forests is still taking place for agriculture, manufacturing, and commercial and housing developments. The current rate of mangrove loss is 1% year^{-1} (Ratanasermpong et al. 2000). Possible collapse of production from the heavy losses of mangrove may occur, and government management plans specify that a time frame must be ascertained to set upper limits for continued exploitation of mangroves and the adjacent terrestrial forests.

The simple mass balance of carbon for Sawi Bay (Table 6.4) indicates that the ecosystem currently has more inputs than losses of carbon. Inputs exceed losses of carbon for a net ecosystem production of ≈112.6 mol C m^{-2} year^{-1}. Most of this excess carbon is probably accumulating in biomass of replanted trees, as most mangrove forests lining the bay are ≤15 years old (Alongi et al. 2000c). Mangroves are the major source of energy and material flow within the bay. The mass balance also indicates that respiration from mangroves and microbes in bay waters is the greatest loss of carbon. Pelagic respiration is a relatively large loss term compared with other tropical coastal ecosystems because of the large export of metabolically active, microbial consortia from aquaculture ponds bordering the bay (Ayukai and Alongi 2000). This suggests that the bay would be sensitive to losses of mangrove forest and to increases in pond effluent.

Assuming a continuation in the decline of mangroves at 1% year^{-1}, a simulation of the changes in the mass balance equation shows that the balance of carbon flow in Sawi Bay changes from positive net ecosystem production to a carbon deficit (NEP < 0) within ≈27 years (Fig. 6.11). This is likely the *maximum* estimate of time until the ecosystem lapses into unsustainability. This iterative change in the mass balance neither considers increases in carbon inputs from erosion of soils, sewage, or pollutants nor the impact of growth of aquaculture ponds with corresponding

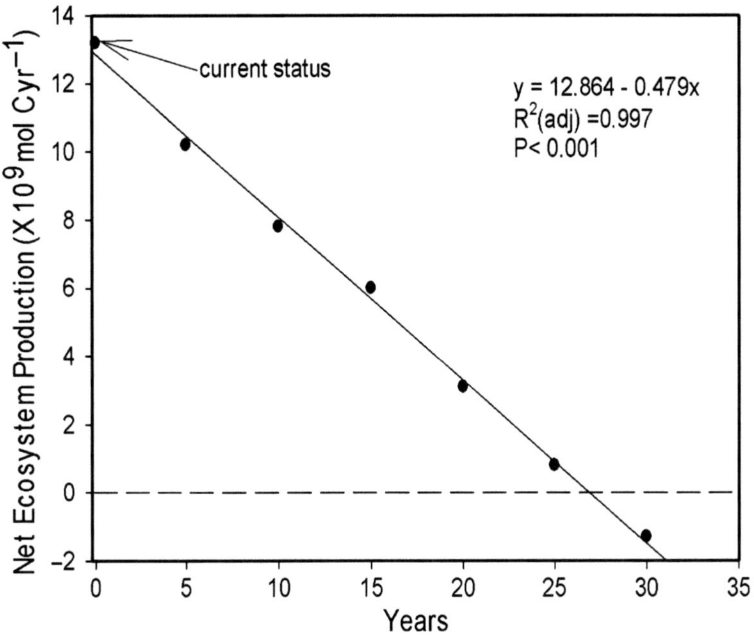

Fig. 6.11 Simulation model predicts a collapse of net ecosystem production of Sawi Bay in southern Thailand, assuming continuation of a 1% annual loss rate of mangroves. Dotted horizontal line represents the threshold of sustainability at NEP = 0 (Updated from Alongi 2005b)

increases in effluent discharge. The model also does not include complex linkages and feedback loops within the ecosystem.

The time boundary is constrained to 27 ± 10 years because the average coefficient of variation of the individual carbon measurements used in the model is ≈33%. Despite this limitation, this estimate has immediate implications for management. Felling trees without or with only limited replanting may lead to ecosystem collapse in as early as 17 years or as late as 37 years. The precautionary principle and the high certainty of both positive and negative feedback loops suggests that management plans to counter the net loss of mangroves should ensure that current rates of replanting equal or exceed current loss rates. An ecosystem that loses more carbon (and energy) than it gains can sustain itself for a limited period of time because of temporal and spatial lags and patchiness between production and consumption processes, but such an ecosystem cannot harvested indefinitely, not without restoring steady-state equilibrium (Odum 1969; Schultze 1994).

6.7.2.2 Mangrove Harvesting Limits and Ecosystem Support

The exemplar of sustainable harvesting of mangroves is the Matang Mangrove Forest Reserve in the state of Perak in peninsular Malaysia. The reserve faces the

Straits of Malacca forming a continuous 48 km belt of mangrove forest with a total area of 48,804 ha, of which 32,746 ha is classified as productive forest. The ecosystem is composed of five major estuaries, lined mostly by *Rhizophora apiculata* with an occasional understory of *Bruguiera parviflora* and *R. mucronata* along river banks (Watson 1928; Muda and Mustafa 2003). Nearly 3,200 ha of forest is conserved for research, tourism, education, a bird sanctuary, and seed stands. There is some fish cage and cockle aquaculture and fishing activity within the reserve, but only non-destructive practices are permitted.

The Matang Reserve was first gazetted in 1902 with the first working plan for sustainable harvesting and conservation completed in 1908. Under the current working plan, productive forest is managed on a sustainable basis using a 30-year rotation cycle, with two thinnings between 15–19 years and 20–24 years, with final felling at 30 years (Muda and Mustafa 2003). The thinned wood is used for poles in housing construction. The area of harvestable *R. apiculata* forest is equally proportioned between the 1- to 10-year, 11- to –20-year and 21- to 30-year-old age classes. Each year, 1,048 ha of forest is harvested in small lots for an average yield of 171 t ha^{-1} per 10 years rotation (Muda and Mustafa 2003). The annual total wood yield currently averages 17,920 t DW. The slash is left to decompose naturally or to be flushed with the tides. When felling lots, a buffer zone of 5–10 m is retained to protect against shoreline erosion. One-half of the lots are allowed to regenerate naturally, although recent problems with inadequate natural regeneration and infestation by *Acrostichum* ferns have necessitated a shift to manual planting and development of seedling banks.

Use of the wood resources in the production of charcoal is the mainstay of the local economy and management of the Matang Reserve. The forests are tendered by the Perak State Forestry Department to charcoal contactors, the number varying from 50–75, each with an average of four to five charcoal kilns. The break-even price for charcoal is about RM 400 t^{-1}; most of the produce is consumed within Perak while some is sent to market in the states of Selangor, Penang, and Kedah. Total net revenue in 2000 was RM 745,300. Net revenue fluctuates, but the total annual forest yield per hectare has remained fairly constant since first harvesting in 1906. This has been attributed to the success of maintaining and managing a balance of 20% reserve forest and 80% harvestable stands.

The mass balance (Table 6.4) shows that mangrove production and respiration dominates carbon flow within the ecosystem. Mangrove production accounts for 95% of total carbon inputs; estimated river/ocean inputs, phytoplankton production and inputs of trash fish to fish cage aquaculture are minor carbon sources. Tree respiration accounts for ≈79% of total ecosystem losses, and all respiration losses account for nearly 86% of total carbon output. Harvesting of timber and poles from thinning stages constitute relatively small losses (1% of total output), as do harvesting of natural and cultured finfish and shellfish within the reserve (5% of total output). Export of mangrove carbon is also a small loss compared with tree respiration, equating to about 5% of total carbon inputs and 24% of mangrove net primary production. Only 2% of TOC input is buried in sediments.

The reserve is producing more carbon than it is losing, with most carbon being stored in new forest growth on accreting mudbanks. Based on extensive surveys, Muda and Mustafa (2003) estimate that the reserve has gained ≈1,500 ha of new forest through accretion since its inception in 1908 (Watson 1928). This would only account for 5% of the excess carbon, but carbon may be retained as peat and as litter in some areas of the reserve. As natural and some managed tropical terrestrial forests in South America have a century-scale capacity to store carbon in wood and roots (Chambers et al. 2000), perhaps mangroves in this reserve have a similar capacity for long-term carbon storage.

Harvesting of 17.1 t DW mangrove wood year^{-1} on 1,048 ha has been sustainable for nearly a century, so assuming that the yield per unit area remains constant, we can model the change in net ecosystem production with possible incremental increases in the areal rate of wood harvesting (Fig. 6.12) using a simple computer simulation of changes to the mass balance equation. An increase in the harvest area at an incremental rate of 50% of the current rate of harvesting leads to a linear decline in net ecosystem production, assuming no complex feedback loops (Fig. 6.12). Solving the model for y = 0, the simulation predicts that 6,261 ha of

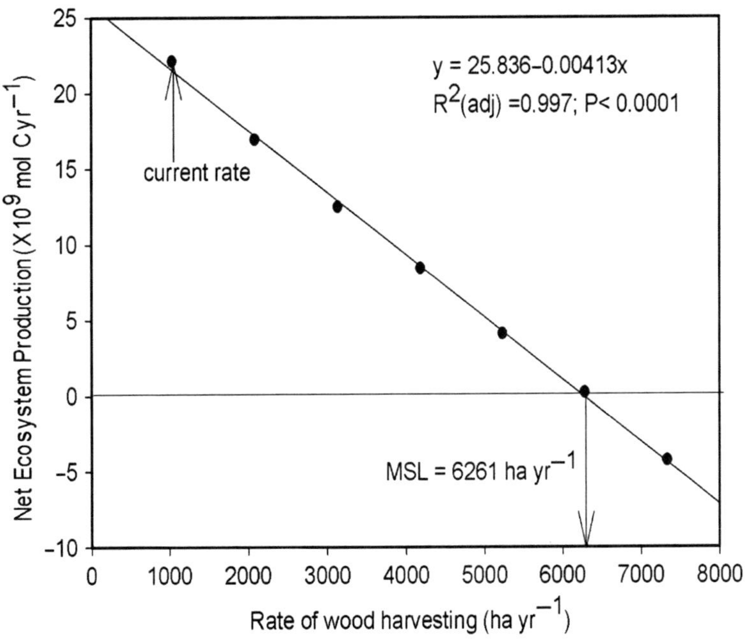

Fig. 6.12 Simulation model predicts a decline in net ecosystem production with incremental increase in rate of timber harvesting in the Matang Mangrove Forest Reserve, Malaysia. Horizontal line represents the threshold of sustainability at NEP = 0. Model run shows that a NEP of zero is attained at a rate of timber harvesting of 6,261 ha^{-1} year (Updated from Alongi 2005b)

mangrove forest year^{-1} would need to be harvested before net ecosystem production is zero and the ecosystem becomes unsustainable. This extraction rate is nearly six times the current rate of harvesting. However, the average coefficient of variation for individual flux measurements is ≈33%, so this translates into a predicted maximum sustainable harvest range of 4,195–8,327 ha year^{-1}. We can cautiously estimate that the current rate of harvesting can be doubled, if necessary, from 1,048 to 2,096 ha year^{-1}. These estimate does not include possible changes in within-stand yield and other changes such as an increase in fish cage aquaculture, but the model does support the fact that the current rate of harvesting is sustainable over the long term and can be increased, if desired.

Sustainable harvesting must involve some understanding of the level of ecosystem support required to service the growth and maintenance of the harvested stands. Ecosystem support is defined here as the total ecosystem area necessary to support the area under extractable use. This idea is based on the assumption that every individual organism is supported, either directly or indirectly, by various biotic and abiotic processes from the adjacent environment. This is reminiscent of the logic used to define the term 'ecological footprint' (Chambers et al. 2000). Here we assume that the area of forest under harvest is supported by the entire coastal ecosystem with the Matang Reserve comprising 8,653 ha (Muda and Mustafa 2003). Currently, the level of ecosystem support is 46.6 ha of total ecosystem area for each hectare of forest harvested (Fig. 6.13). With an incremental

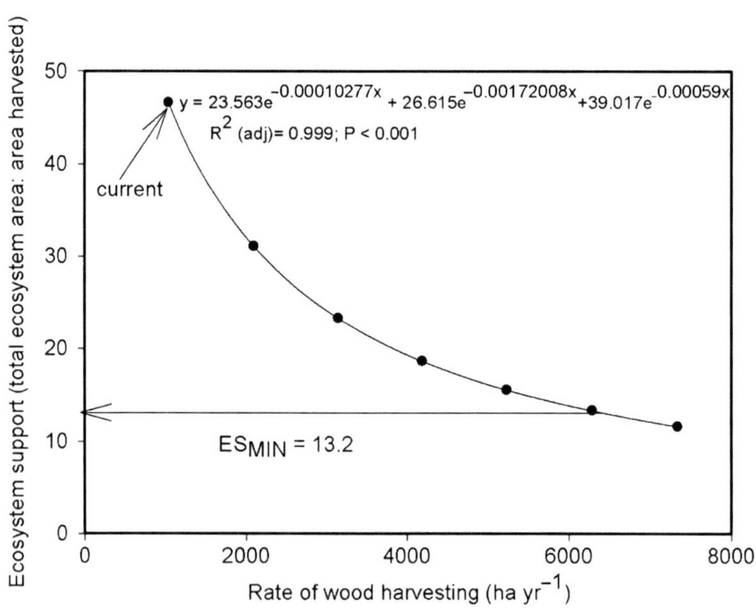

Fig. 6.13 Simulation model predicts an exponential decline in ecosystem support with incremental increase in timber harvesting within the Matang Mangrove Forest Reserve, Malaysia (Updated from Alongi 2005b)

decline in net ecosystem production as the harvested area increases, the level of ecosystem support undergoes an exponential decline when we iterate the changes in the mass balance equation in a simple computer simulation. At the predicted average maximum rate of harvesting (6,261 ha year^{-1}), when net ecosystem production is zero, the level of ecosystem support is 13.2 (Fig. 6.13). At the recommended rate of harvesting of 2,096 ha year^{-1}, the level of ecosystem support is 31 (Fig. 6.13). If we calculate changes in ecosystem support on the basis of wood yield, the index is calculated by dividing the total ecosystem area by total yield with the specific value having units of hectares per tonnes (Fig. 6.14). If the predicted maximum rate of harvesting is used, each tonne of wood harvested requires 0.46 ha of total ecosystem area for support (Fig. 6.14). At the current harvesting rate and at the recommended doubling of the harvesting rate, the level of ecosystem support required to sustain each tonne of wood harvested is 2.8 and 1.4 ha, respectively (Fig. 16.14). Similar calculations for terrestrial forests indicate a range of values of 1.0–5.7 ha t^{-1} (Chambers et al. 2000) suggesting that the ecological dynamics of mangrove harvesting is similar to that of other forested ecosystems.

As noted earlier, few studies have attempted to estimate the level of mangrove ecosystem area required to support sustainable aquaculture (Robertson and Phillips 1995; Larsson and Padilla 1999). An estimate by Robertson and Phillips (1995) of the impact of shrimp pond effluent on mangroves indicates that if the effluent is delivered directly into forests, 2–22 ha of forest is required to totally utilize the

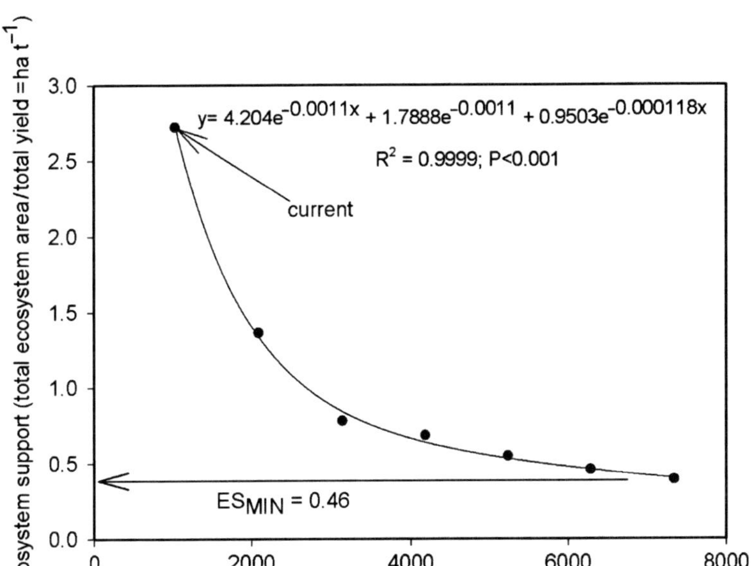

Fig. 6.14 Simulation model predicts an exponential decline in ecosystem support with an incremental increase in timber harvesting within the Matang Mangrove Forest Reserve, Malaysia (Updated from Alongi 2005b)

nitrogen and phosphorus loads from a 1 ha shrimp pond. Semi-intensive ponds require less area (2–3:1) than intensive ponds which are estimated to require an areal ratio of 7:1 to process N and an areal ratio of 22:1 to assimilate P. In a study of semi-intensive shrimp farming in Colombia, Larsson et al. (1994) estimates that 35–190 ha of mangrove ecosystem is needed to support each ha of pond, as more than 80% of the primary production required to feed shrimp stock is derived from adjacent ecosystems. Aquaculture within mangrove ecosystems ranks as one of the most resource-intensive systems and can be characterized as unsustainable.

The mass balance estimates of the exploitation of mangrove forests in Thailand and Malaysia are not absolute, but are intended to show that comparatively simple models based on empirical data can quantify estimates of uppers limits of sustainable use of resources. Practical management needs relatively simple and practical approaches for approximating upper limits to sustainable production and ecosystem support. Mass balance models cannot be used in isolation from other information in order to set management plans; they do have limitations. For instance, the iterative adjustments to the mass balance do not take into account linkages and feedbacks among ecosystem components. It is almost certain that a change in the rate of harvesting would affect detrital outputs (e.g., more slash on the forest floor) that would in turn affect the rate of carbon loss via microbial respiration, and so forth. More complex models would have to refine these estimates to account for such interactions.

Regardless of the accuracy of the mass balance calculations, the first example from southern Thailand underscores the consequences of unregulated or poorly managed mangrove cultivation and felling. The Malaysian example illustrates best-practice methods in sustainable extraction of mangrove timber and offers a valuable lesson for estimating limits to sustainability of other *Rhizophora apiculata* plantations. The level of ecosystem support for *Rhizophora apiculata* in the Malaysian scenario is 2.76 ha t^{-1}. This estimate is well within the range of 1–5 ha t^{-1} estimated for terrestrial plantations (Chambers et al. 2000). This similarity suggests that methods and practices used to manage the harvesting of terrestrial trees of a similar age span and with similar growth rates can be applied to *Rhizophora apiculata* plantations. As illustrated here, many functions of mangrove forest ecosystems appear to be more similar to their terrestrial forest counterparts than their aquatic neighbors.

Chapter 7
Synthesis

7.1 Developing a Global View

7.1.1 A Budget and Its Implications

Since the 1990s a number of attempts have been made to place the functional importance of mangroves within a global context (Twilley et al. 1992; Saenger and Snedaker 1993; Gattuso et al. 1998; Jennerjahn and Ittekkot 2002; Borges 2005; Duarte et al. 2005; Dittmar et al. 2006; Alongi 2007; Bouillon et al. 2007a). Based on average fluxes of individual carbon processes as calculated within this book, I summarize our understanding to date of the major pathways of carbon flow through the world's mangrove ecosystems (Fig. 7.1). As with all budgets, the model is indicative, not absolute, and hopefully instructive in pinpointing patterns of flow important relative to global carbon cycling and where further research should be focused. Processes such as secondary production and benthic primary production are not included in the budget.

The largest flux of carbon in mangrove ecosystems is between the trees and atmosphere; a little more than half of gross primary production is respired by the trees. Net forest primary production is accounted for by litterfall and wood production (each ≈20% of NPP; these values are identical to those calculated by Bouillon et al. (2007a). By difference, fine root production equates to roughly 50% of NPP. This latter value is highly subjective; we have earlier noted that there are exceedingly few measurements of fine root production for mangrove trees (Section 2.5.2), but it does highlight that below-ground production of roots is a major process, and that more empirical estimates are urgently needed. Adding canopy respiration, soil respiration, and pelagic respiration, total ecosystem respiration is 500 TgC year^{-1}, which equates to nearly 70% of forest GPP. Carbon burial averages 29 TgC year^{-1} which is about 10% of forest NPP, and export of POC and DOC equates to 43 TgC year^{-1} which is about 10–15% of forest NPP.

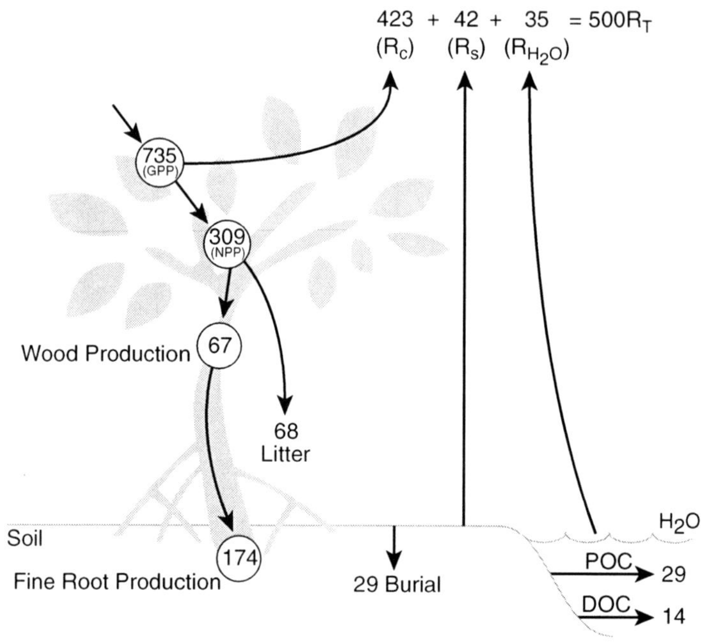

Fig. 7.1 Model of the major pathways of carbon flow through the world's mangrove ecosystems. Units = TgC year^{-1}. The budget assumes a global ecosystem area of 160,000 ha

A net ecosystem production of ≈160 TgC year^{-1} is estimated by subtracting carbon losses from gains. What does this figure really mean, considering that all of the major inputs and outputs are included, especially net tree growth? Bouillon et al. (2007a) calculated a similar figure (112 TgC year^{-1}) as "unaccounted for" mangrove production and suggested that this value may represent possible DIC export to the coastal ocean, assuming that the forests are in steady-state. This is a reasonable idea because, as discussed in Section 6.3.2, much of the inorganic carbon derived from organic carbon mineralization in the soil is probably transported laterally from the forest floor. Such unusual fluxes are too often overlooked in biogeochemical studies of intertidal ecosystems. Many of my own studies (e.g., Alongi 2001) indicate that substantial amounts of inorganic carbon are unaccounted for; actual soil mineralization rates are likely to be higher than gleaned from CO_2 measurements taken only across the surface soil. A few facts suggest that this idea is tenable: (1) there is significant microbial activity to soil depths of >1 m, (2) shoreline morphology and visual observations over tidal cycles indicate significant lateral drainage of porewater during receding tides, and (3) the sum of individual carbon metabolism measurements is often greater than the rate of total carbon metabolism measured from the soil surface; this is especially so in macrotidal environments (e.g., Alongi 2001). In these latter systems, the geomorphology is such that mangroves sit atop highly sloped escarpments from which a large quantity of interstitial water can alternatively be drained and replenished (Woodroffe 2003).

In addition to the low number of empirical estimates of many processes and their high variability, other processes not included in the budget can constrain the amount of 'unaccounted for' net mangrove production. These processes include faunal production and respiration, virus repellents, protective metabolites, mutualism, and symbiosis, or more subtle mechanisms that require use of fixed carbon. In other aquatic ecosystems, many organisms apparently use surplus carbon for purposes that promote fitness, but there are genuine cases where carbon is in excess and is disposed of either in organic or inorganic form (Hessen and Anderson 2008). Such may be the case for mangrove ecosystems. Indeed, if the large production of fine roots (174 TgC year^{-1}) is correct, rapid turnover/metabolism of these fine roots would equal much of our excess carbon estimate ($\approx$160 TgC year^{-1}).

From an evolutionary perspective, it makes sense for mangroves to invest a large proportion of growth in roots as, unlike terrestrial forests, carbon is lost via lateral and advective transport in the form of litter and soil interstitial water. Proportionally more carbon allocated below-ground would be an effective counterbalance to such losses that terrestrial forests do not experience. The principal pathway to recycle nutrients in a tropical terrestrial forest is via the rapid recycling of foliage (litter) in a relatively thin layer of soil humus, whereas in mangrove forests the main pathway is via close-knit cycling between roots and microbes to deep into the soil where tidal drainage still exerts a powerful force, although less so than on the surface of the forest floor. Further, large below-ground pools of dead roots often found in mangrove forests may be a critically important investment in recycling pathways to curtail carbon loss via tides. Clearly, life in such a highly dynamic environment has its energetic costs.

7.1.2 *The Contribution of Mangroves to Carbon Cycling in the Global Coastal Ocean*

Assuming that the global carbon budget illustrated in Fig. 7.1 is reasonable, to what extent does the world's mangrove ecosystems contribute to carbon cycling in the global coastal ocean? Let's first look at the mangrove contribution to carbon

Table 7.1 The global contribution of mangroves and other coastal habitats to carbon burial in the global coastal ocean (Updated from Duarte et al. 2005)

Component	Area (10^{12} m^2)	Burial rate (gC m^{-2} year^{-1})	Global C Burial (TgC year^{-1})
Mangroves	0.16 (0.5%)	181.3	29.0 (12%)
Salt Marsh	0.4 (1.4%)	151.0	60.4 (25%)
Seagrass	0.3 (1%)	83.0	27.4 (11%)
Estuaries	1.8 (6.2%)	45.0	81.0 (33%)
Shelf	26.6 (90.9%)	17.0	45.2 (19%)
Total			243.0

burial (Table 7.1). Updating the data in Table 3 from Duarte et al. (2005), we find that all of the vegetated and estuarine habitats contribute to carbon burial in the coastal sediments disproportionate to their relative areas. However, carbon burial in mangrove soil is 24 times greater in proportion to habitat area as compared with salt marshes (18 times), seagrass beds (11 times) and unvegetated estuarine sediments (5 times); by comparison, proportionally little carbon is buried on the world's continental shelves. Mangrove peat found in many areas is evidence of their tendency to store carbon below-ground. Until replaced by terrestrial forests, carbon accumulating of mangrove wood may also be considered a mode of storage. As in terrestrial forests, there may be significant storage in mangrove tree biomass, especially if the forests are long-lived.

How do mangroves compare in terms of metabolic balance? Again, updating Table 3 in Duarte et al. (2005) we find that although mangroves occupy only about 0.5% of total coastal area globally, their rates of GPP, NEP, and R account for ≈5–6% of the global total, clearly contributing a small, but disproportionate, share to global carbon cycling (Table 7.2).

How do mangroves compare with other tropical forests in terms of carbon flow? Compared with tropical humid evergreen forests, mangroves vest more fixation in fine root production than in foliage, and to a lesser extent, wood (Table 7.3). This makes functional and evolutionary sense as discussed earlier, considering that mangroves are subjected to tides and live in anaerobic soils. It is puzzling then why heterotrophic respiration in mangroves is low compared with terrestrial forests (Perry et al. 2008). This is not true, however, if most remaining carbon attributed as net ecosystem production is actually microbial carbon respiration exported from soils (the 'unaccounted for' carbon discussed earlier). If this idea was true, then the rates of heterotrophic respiration and the ratio of R_e/GPP are equivalent between forest types (0.90 for mangroves assuming the NEP as the missing DIC vs 0.88 for terrestrial forests). However, mangrove NPP is still a greater proportion of GPP than for their terrestrial counterparts. Given the high variability within each set of values, it seems clear that mangroves and tropical terrestrial forests allocate carbon similarly.

Table 7.2 The global contribution of mangroves and other coastal habitats to carbon metabolism in the global coastal ocean (Updated from Duarte et al. 2005)

Component	Area (10^{12} m^2)	R	Global R	GPP	Global GPP	NEP	Global NEP
Mangroves	0.16	3,125	500	4,594	735	1,012.5	163
Salt Marsh	0.4	2,010	804	3,595	1,438	1,585	634
Seagrass	0.3	692	228	1,903	628	1,211	400
Macroalgae	1.4	2,116	2,962	3,702	5,183	1,587	2,221
Coral Reefs	0.6	1,572	943	1,720	1,032	148	84
Unvegetated sediments	23.9	83	1,992	68	1,624	−15	−370
Global benthic coastal ocean	26.76		7,429		10,640		3,132
% Mangroves	0.6%		6.7%		6.9%		5.2%

Table 7.3 Comparison of the major carbon fluxes (g C m^{-2} $year^{-1}$) through mangrove and other tropical humid forests (Terrestrial data from Litton et al. 2007 and Luyssaert et al. 2007)

Process	Mangrove forests	Tropical humid forests
GPP	4,596	3,551
NPP (% of GPP)	1,930 (42%)	852 (24%)
fNPP (% of total NPP)	425 (22%)	316 (37%)
wNPP	419 (22%)	212 (25%)
rNPP	1,086 (56%)	324 (38%)
NEP	1,018	403
Re	3,125	3,061
Ra	2,644	2,323
Rh	481	877
Re/GPP	0.68	0.88
	0.90[a]	

[a] Assumes NEP as DIC export.
Abbreviations: GPP = gross primary production, NPP = net primary production, fNPP = foliage net primary production, wNPP = wood net primary production, eNPP = root net primary production, NEP = net ecosystem production, Re = total ecosystem respiration, Ra = canopy respiration, Rh = heterotrophic respiration.

7.2 The Most Important Facts Regarding Mangrove Energetics

We conclude with a list of the most important facts pertinent to the energetics of mangroves:

- Mean above-ground biomass of mangrove forests averages 247.4 t ha^{-1}, which is equal to the global average for tropical terrestrial forests.
- Most nutrient elements are stored in soils and secondarily in dead roots. Large pools of dead roots serve to store and conserve nutrients below-ground.
- Mangroves follow a conservative water-use strategy with low transpiration rates and stomatal conductances and high water-use efficiencies compared with other C_3 plants. There are functional tradeoffs between retaining water and gaining CO_2 to minimize water-use and maximize photosynthesis.
- Maximum photosynthetic rates can exceed 25 μmol CO_2 m^{-2} s^{-1} but most lie between 5–20 μmol CO_2 m^{-2} s^{-1} with photosynthetic saturation at comparatively low light levels. Shade-intolerant terrestrial trees have a median photosynthetic rate (13 μmol CO_2 m^{-2} s^{-1}), slightly higher than the mangrove median of 12 μmol CO_2 m^{-2} s^{-1}.
- Rates of dark respiration in mangrove leaves range from 0.2 –2.0 μmol CO_2 m^{-2} s^{-1} with photosynthesis to respiration ratios ranging from 3.4– 12.2, which are at the upper end of the range for tropical terrestrial trees.
- Across all species and ages, mean annual growth increment (at diameter-at-breast height) of mangrove stems range from 0.1–1.8 cm $year^{-1}$.

- The mean rate of above-ground net primary production of mangroves is 11.13 t DW ha^{-1} year^{-1} (=44.52 mol C m^{-2} year^{-1} assuming 48% C content of dry wood). For terrestrial forests, the mean rate of above-ground NPP is 11.93 t DW ha^{-1} year^{-1}. Rates of mangrove above-ground NPP are roughly equivalent to tropical terrestrial forests.
- Mangroves are usually either N- or P-limited, or both, depending upon position in the intertidal, species composition, extent of terrigenous input, soil fertility and texture, soil redox status, and salinity.
- Compared with leaf resorption in other forests, nitrogen is resorbed in mangrove leaves at efficiencies at the higher end of the range; efficiency of P resorption in mangrove foliage is within the mid-range compared with terrestrial forests.
- In well-developed canopies, algal production is dwarfed by tree production due to severe light limitation. In open canopies, and especially in polluted systems, algal production is proportionally greater.
- Links between mangrove trees and animals in the canopy and on roots above-ground are extremely diverse, with close associations between tree and animal pollinators and mutual interactions of mangroves with ants, butterflies, monkeys, and birds. The actual impact of these interactions on canopy production is poorly understood.
- Insect herbivory accounts for only a small to modest percentage loss of canopy biomass (usually <10%), but there are documented cases of defoliation of entire stands.
- Root epibiontic communities thrive in some forests, especially when tidal waters are sufficiently transparent to permit algal photosynthesis. These variegated and highly diverse communities can attain a high biomass on some prop roots and facilitate root growth and production, but little work has been done on the significance of these communities to ecosystem energetics.
- Tidal and wave energy constitute an auxiliary energy flow or energy subsidy in mangrove estuaries. Tidal range varies greatly worldwide, but tidal circulation within most mangrove waterways is characterized by a pronounced asymmetry between the stronger ebb and weaker flood tides. This asymmetry results in self-scouring of the waterways. There are natural feedbacks between vegetation, water flow, and sedimentation. These feedback mechanisms have great implications for human impacts on mangroves.
- Lateral trapping of water within the forest is a dominant process controlling longitudinal mixing in mangrove waterways. The trapping phenomenon occurs when some of the water flowing in and out of an estuary is temporarily retained in the forest to be returned to the main water channel later. Mangrove estuaries exhibit secondary circulation patterns superimposed on the primary tidal circulation. This phenomenon is responsible for the often observed trapping of floating mangrove debris, including propagules, in density-driven convergence fronts.
- The presence of trees, roots, animal burrows and mounds, timber and other decaying vegetation lying on the forest floor, exert a drag on the movement of tidal waters within forests. The drag force has two main influences: (1) inundation of the forest is inhibited and (2) trapping of water in the forest is enhanced.

- Crabs and other benthic organisms produce numerous burrows and other biogenic structures in the forest floor. The total quantity of water that flows through burrows in a 1 km^2 area of forest ranges from 1,000 to 10,000 m^3, representing from 0.3% to 3% of the water volume moving through the forest. Well-flushed burrows are therefore an efficient mechanism by which salt can be transported away from roots. Diffusion of salt to a burrow will be more efficient and faster than diffusion to the soil surface.
- Mangrove forests often experience significant groundwater flow. This flow can be an adaptive advantage in that groundwater flow is likely to be an important pathway for the removal of salt excluded by tree roots and reduced by-products of microbial decomposition. It is likely that crab burrows, fissures, and cracks in the substrate facilitate the movement of groundwater.
- The influence of groundwater-derived nutrients on nutrient dynamics in mangrove waterways can be seasonally significant. Groundwater seepages may promote the growth of mangroves and associated flora in areas where growth may be otherwise unsuitable, by reducing anoxia and providing an crucial input of freshwater and dilution of salt.
- Mangroves often receive a substantial amount of wave action, especially when growing on the fringes of open bays and estuaries. Mangrove forests attenuate wave energy via multiple interactions of waves with mangrove trunks and roots and by bottom friction. Mangroves are effective at protecting coastlines from erosion.
- Tidal flow and subsequent attenuation of water flow results in the deposition of fine sediments. The transport of suspended sediment in mangroves is controlled by a number of inter-related processes.
- Mass accumulation rates of sediments range from <1 mm $year^{-1}$ to more than a few centimeters per year, highest in mangroves lining rivers with high freshwater discharge rates or in mangroves within highly impacted river catchments. Lowest accumulation rates occur in fringing mangroves bordering open bays and estuaries in dry tropical regions.
- The development of mangroves and shorelines is closely interlinked. Colonization by propagules takes place when a mudbank forms and stabilizes. Over time, these pioneers grow into a young forest; stand development takes place in which competition for space and light leads to changes in forest structure and composition, modified by gap processes in which local mortality gives way to colonization of gaps. An erosion phase may occur followed by an accretion phase; sediments that are remobilized by erosion are transported and eventually re-deposited in an area where tides and currents permit settlement and stabilization of a new mud bank, completing the cycle of mangrove and shoreline development.
- Mangrove waterways are usually very turbid due to self-scouring, and transport and suspension of fine particles. One often-recurring pattern is a link between pH and oxygen and DOM concentrations, especially polyphenolic compounds. Most DOC is sourced from the leaching of mangrove leaves which contain high concentrations of tannins and other phenolic compounds to ward off herbivores. These compounds undergo rapid transformation by microbes, photooxidation,

and physical processes, having an impact on the subsequent bioavailability of DOM. Polyphenolic compounds react and bind to sugars, proteins, lipids, and enzymes, making these labile substances less available for biological use.

- Archaea, bacteria, protists, and viruses occupy a central role in the energetics of pelagic food webs in mangroves. A large proportion of DOM and energy flow is funneled through these highly diverse, actively growing, assemblages – many of mixed trophic states – and subsequently transferred to higher consumers via a chain of small protistan grazers in what essentially is a microbial loop, the hub of the entire pelagic food web. The study of the energetics of these consortia has lagged behind the study of energetic processes within the forest.
- Rates of bacterioplankton productivity in relatively pristine mangrove waters range from 0.1–22.0 $\mu g\ C\ l^{-1}\ h^{-1}$ and from 10–91 $\mu g\ C\ l^{-1}\ h^{-1}$ in eutrophic waters. These rates are greater than those measured in salt marsh waterways (4–5 $\mu g\ C\ l^{-1}\ h^{-1}$). Specific growth rates and rates of bacterioplankton productivity mirror rates of phytoplankton production, reflecting a key link in the microbial loop. The ratio of bacterioplankton to phytoplankton production is high (mean = 106%) compared to other estuarine waters implying that other sources of organic matter (e.g., mangrove detritus) are needed to sustain bacterial growth.
- Grazing experiments point to intense consumption of mangrove bacterioplankton by microzooplankton.
- Phytoplankton assemblages are often of low species diversity. In pristine mangrove estuaries, phytoplankton production typically ranges from 20–500 $mg\ C\ m^{-2}\ day^{-1}$ and standing stocks of chlorophyll *a* are usually within the range of <1–3.5 μg chl *a* l^{-1}. Standing stocks of phytoplankton are low, but phytoplankton production in mangrove waters is at the mid- to high-end of the range of rates measured in salt marshes and temperate estuaries. Phytoplankton production is regulated by light and to a lesser extent, temperature, nutrients, herbivory, and physical processes. There is some evidence for phosphorus rather than nitrogen limitation.
- Mangrove waters are net autotrophic, with mean P_G/R of 1.8, as rates of pelagic respiration average 846.9 $mg\ C\ m^{-2}\ day^{-1}$ and range from 0.1–3.5 $g\ C\ m^{-2}\ day^{-1}$. This mean rate is less than the mean of 1,368 $mg\ C\ m^{-2}\ day^{-1}$ calculated for temperate and sub-tropical estuaries. A significant, but weak, relationship was found between chlorophyll *a* concentrations and rates of respiration implying a link between microheterotrophs and phytoplankton, but it may also indicate that phytoplankton contribute greatly to water-column respiration.
- Numbers and biomass of mangrove zooplankton vary widely, but are usually greater than the size of zooplankton communities in adjacent coastal waters. The most important factors regulating changes in mangrove zooplankton communities are salinity and tides. Most zooplankton studies have focused on community structure rather than population or community energetics. Secondary production has been measured for a few species of mangrove zooplankton. Egg production rates of some dominant copepods range from 0.8–51.4 eggs $female^{-1}\ day^{-1}$ – at the high end of the range of rates measured in other marine environments. Production rates are less in dry than in wet tropical systems. Factors other than temperature (e.g., food availability) may regulate egg production.

- Studies of zooplankton diets have focused on juvenile stages of shrimps. Recent observations confirm that phytoplankton and mangrove-derived organic matter is ingested within mangroves, but the mangrove signal rapidly decreases with distance from the forests. Some zooplankton feed discriminately, selecting phytoplankton cells among the suspended matter in creeks and waterways. Mysids feed mainly on phytoplankton, ciliates, and flagellates, with no evidence that detritus is consumed. Seasonality and other external factors (e.g., excess nutrient loads) also play an important role in trophic relationships among the plankton. Laboratory experiments confirm that calanoid and cyclopoid copopods feed on algae, but also show that fecal material from grapsid crabs when mixed with algae results in greater assimilation rates, thus linking planktonic and benthic food webs.
- Penaeid shrimps constitute the bulk of the commercially important crustaceans taken from mangrove estuaries. Higher densities of juvenile penaeids are found in mangrove nursery areas than in adjacent habitats. Penaeids eat diatoms, filamentous and epiphytic algae, microbial-detritial aggregates, foraminifera, and a variety of other organisms. These prey obtain carbon and nitrogen from benthic algae and/or phytoplankton and, to a lesser extent, from mangroves. The extent of the relative contribution of algal versus mangrove carbon and nitrogen appears to be dependent on distance from mangrove estuaries.
- Rates of annual shrimp production are at the high end of the range of values (13–756 kg ha^{-1} $year^{-1}$) for shrimps globally. Annual landings of shrimp within mangrove waters worldwide quadrupled from <100,000 t in 1950 to nearly 300,000 t in 1980, with a decline to ≈200,000 t in 2006, indicating overexploitation.
- Numbers of fish species range from <10 to nearly 200 in any particular estuary with a tendency for more species in larger systems. Fish densities and biomass vary widely, but are greater than in temperate estuaries. Recent work has clarified food resources previously defined as 'detritus', 'indistinguishable', or 'amorphous' material. Much of the amorphous material in fish guts is seagrass detritus. Juvenile zooplanktivorous and omnivorous fish prefer crustaceans, while fishes and algae are the preferred foods of piscivores and herbivores. Several species feed extensively on sesarmid crabs thereby 'short-circuiting' the traditional mangrove food web. Living in these 'interconnected habitat mosaics' limits the level of direct dependence of fish and crustaceans on mangroves. This crucial but tenous link is unlikely to surface in catch data and statistical analyses.
- Fish productivity in mangrove estuaries ranges from 17–1,646 kg ha^{-1} $year^{-1}$ with most estimates commonly within a few hundred kilograms per hectare per year, somewhat greater than penaeid shrimp production.
- The forest floor can vary from dead hard corals and boulders to sand and mud. In mature forests, soils contain significant quantities of peat and fine fibrous roots whereas in forests fringing river banks, soils often possess large quantities of sand. Forest age may be an important factor in the maturation of soils (e.g., accumulation of organic matter, buildup of dead roots). Soil organic content and the C/N ratio tends to increase from pioneering to mature mangroves, as in terrestrial forests. Soil organic matter is composed of mangrove litter, roots,

wood, and mangrove peat, interspersed with material derived from seagrasses and other halophytes, up-river terrestrial soils and plants, animal wastes, microbial biomass, and organic marine material.

- Benthic communities are tightly interlinked with the trees, exhibiting complex patterns of zonation across intertidal zones and with height above the forest floor. Gastropods and crustaceans are the major invertebrate groups living on the soil surface. Their patterns of distribution and abundance relate to frequency of tidal inundation, competition, food availability, and predation. Many studies have examined the reproductive histories of mangrove benthos, but estimates of secondary production are sparse.
- A large proportion of litter reaching the forest floor is consumed or hidden below-ground by grapsid and ocypodid crabs, thereby reducing the amount of material available for export and serving as another mechanism to retain nutrients. Consumption rates of litter are usually quite high, but not always. Consumption rates vary, depending on the relative frequency of tidal inundation and nutritional quality of the litter, and with mangrove and crab species. Competition between crabs and other benthic consumers also plays a role in the extent of leaf litter consumed. Shredding, ingesting, and assimilating litter plays an important role in facilitating decomposition of the material. Consumption of leaf litter facilitates microbial colonization by producing material, including feces, with a greatly increased surface area to volume ratio, thus acting as a positive feedback loop in the flow of energy and materials within benthic food webs.
- Crabs and other benthic detritivores initially process nearly 50% of the total litterfall produced in many forests. The rest of the material is swept away by the tides but any litter remaining in the forest is then further decomposed by various microbial assemblages. Litter shredded but unassimilated by the crabs is returned to the soil to be consumed by microbes, as is litter not immediately processed by macroconsumers. Several common features emerge in microbial decay of mangrove leaf detritus: (1) absolute decay parameters are site- and species-specific, (2) leaves decompose faster in subtidal than in intertidal areas, (3) leaf decay is more effective when leaves retain their moisture, (4) leaves of species with lower tannin content and higher initial nitrogen concentrations decompose faster, and (5) decomposition of leaves of the same species occurs at similar rates in both the tropics and subtropics, but in the dry tropics, decomposition rates are slowed by seasonal aridity and high salinity. The decay constants of yellow leaves deposited onto mangrove sediment range from 0.001–0.1 day^{-1} with decomposition usually following an exponential pattern.
- Burrowing grapsid and ocypodid crabs dominate the benthic infauna. They are 'ecosystem engineers', crucial in ecosystem functioning by: (1) redistributing fluids, gases, and solids by reworking of soil or by active pumping of material in and out of burrows; (2) handling and ingesting potential food resources and by doing so modifying and changing the state of the soil (e.g., redox); (3) mediating transport of substances via diffusive gas transport and passive fluid transport; and (4) changing the reactive states of substances via removal of toxic metabolites (e.g. hydrogen sulfide) and by introducing oxygen deep into the soil.

- Fatty acid and stable isotope analyses show that many epifauna and infauna lack a distinct mangrove dietary signal; a more common signal suggests that benthic algae, bacteria, and organic matter plus some plankton are consumed. A new paradigm has emerged indicating that most soil-dwelling fauna prefer algal foods. Mangrove detritus is still a key food resource for some organisms, but probably plays a more crucial role in nutrient recycling and retention.
- Mangrove trees eventually die, fall, and come to rest on the forest floor where they are quickly colonized by fungi and other organisms that either live within the forest or are imported with the tide. Teredinid mollusks are the main agents of wood decomposition, possessing symbiotic cellulolytic and nitrogen-fixing bacteria that assist in breaking down the wood. The energetic significance of wood decomposition is unknown but it may be large.
- Few studies have measured either root production or decomposition. The sparse data suggest that below-ground root decomposition is slow (0.06–0.34% day^{-1}) compared with breakdown of other tree components. Slow decay would help explain the formation of peat in many forests. Peat formation is another mechanism to store and retain essential elements.
- High rates of mangrove primary productivity depend not only on unique and highly evolved physiological mechanisms, but also on highly efficient interrelationships with soil microbes. Archaea, bacteria, fungi, and other microflora alter the microenvironment around mangrove roots via their metabolic activities, transforming and releasing nutrients, and modifying soil chemistry. The relationship between microbes and trees is complex, as both trees and microbes share and complete for limiting nutrients. Nutrient-use efficiencies of mangroves are equal to or higher than those of other tropical trees implying that bacterial transformation of nutrients and shunting to trees is rapid. Mangrove-microbe relations have been most closely observed within the rhizosphere where highly specialized groups of bacteria and fungi coexist.
- Measurement of benthic respiration across the soil surface represents decomposition of organic matter in surface and near-surface soils, but it is unlikely to be a good overall measure of total organic matter decomposition within the forest floor, due to losses via lateral transport during ebb tide and high rates of deep, subsurface respiration. Rates of surface oxygen and carbon dioxide flux range from 50.6 ± 2.2 and 60.1 ± 1.5 mmol m^{-2} day^{-1} with higher rates from exposed soils; the mean respiratory quotient is ≈ 1.5.
- Aerobic respiration and anaerobic sulfate reduction are the main decomposition pathways in mangrove soil. As oxygen penetrates and is depleted within a few millimeters, anaerobic metabolism dominates below-ground decomposition. Sulfate reduction averages 36.2 mmol S m^{-2} day^{-1} and ranges from 0.2–319.0 mmol S m^{-2} day^{-1}. On average, the rates of sulfate reduction in mangrove soils appear to be less than rates measured in salt marshes. Sulfate reduction rates decline in *Rhizophora apiculata* forests older than about 35 years, indicating that other metabolic pathways may dominate in older stands. Iron reduction (20.6–63.4 mmol C m^{-2} day^{-1}) may be important, but more work needs to be done in acquiring detailed measurements from a range of forests.

- Methanogenic bacteria account for a small fraction (1–10%) of total microbial carbon decomposition. Methane production is usually low (0.1–5.1 mmol CH_4 m^{-2} day^{-1}) and highly variable and in some forests cannot be detected. In highly polluted mangroves, rates can reach 60 mmol CH_4 m^{-2} day^{-1} and in subtropical mangroves, maximum rates approach 30 mmol m^{-2} day^{-1}. Methanogenesis also occurs in and on parts of the trees, so its overall significance to carbon flux is probably underestimated.
- Mangrove waterways are sources of CH_4, CO_2, and other greenhouse gases. Water-air CO_2 fluxes from mangrove waterways and adjacent inshore waters suggest that mineralization of organic matter and subsequent gas emissions to the atmosphere could represent significant export pathways. CO_2 flux rates average 43–73 mmol C m^{-2} day^{-1}, corresponding to ≈7% of the total emission from oceanic waters at subtropical and tropical latitudes, and about 24% of the total CO_2 emissions from coastal waters, disproportionate to their relatively small area. Tidal waters are supersaturated in CH_4, CO_2, and N_20 compared with atmospheric concentrations. CH_4 emissions range from 3.3–10.4 mmol CH_4 m^{-2} day^{-1} in pristine waters to 5,216 mmol CH_4 m^{-2} day^{-1} in highly polluted mangroves. N_2O emissions range from 2.9–31.2 μmol m^{-2} day^{-1}. Mangroves appear to be a greater source of these gases to the atmosphere than previously believed as emission rates are usually greater from creek water than from the forest.
- Nitrogen budgets for three Indo-West-Pacific forests show that most soil organic nitrogen is ammonified. Algae can be important in the N cycles as most DIN flux across the soil–water interface is taken up by benthic micro- and macroalgae. Little nitrogen is buried in soil. Most NH_4^+ is taken up by tree roots as uptake of dissolved N from the water column only accounts 5–10% of N required for mangrove net primary production. Rates of ammonification are tightly linked to rates of total nitrogen input. The mineralization and burial efficiencies of soil nitrogen do not correlate to the rate of nitrogen input, but the mineralization efficiencies range narrowly from 67–92%. Burial efficiencies exhibit a wider range of 4–31% but burial is most often <15%.
- Denitrification ranges from zero to 11,000 μmol N m^{-2} day^{-1} with a mean denitrification rate of 1,532 μmol N m^{-2} day^{-1}. Measurements are highly method-dependent so these values must be treated cautiously. Denitrification as an N export pathway in mangroves is of proportionally less significance than in other aquatic ecosystems. Denitrification is regulated by nitrate availability, temperature, salinity, and soil organic matter content.
- Denitrification losses are not counterbalanced by gains from nitrogen fixation. In mangrove soils low nitrogen fixation activity is mediated by sulfate reducers and microbes associated with mangrove roots. N_2 fixation ranges from zero-4316 μmol N m^{-2} day^{-1} with a mean rate of 616 μmol N m^{-2} day^{-1}. These rates are less than those for denitrification and are also less than those measured in salt marshes and seagrass beds. However, nitrogen fixers are very active elsewhere in the canopy, such as on prop roots, litter, fresh leaves, bark, logs, and other wood. Nitrogen-fixing microbes that reside in the rhizosphere are now being used to promote mangrove growth for forest restoration.

- Nitrous oxide, an intermediate product of nitrification and denitrification and important in the greenhouse effect, has been measured from mangrove soils with rates varying from undetectable or showing net uptake, to release of up to 330.0 μmol m^{-2} day^{-1}.
- Transformation of phosphorus in mangrove soils is poorly understood. Uptake of soluble P by mangroves involves mutualistic interrelationships between microbes and tree roots. Arbuscular mycorrhizal fungi in the rhizosphere benefit from oxygen translocated by the trees to their roots, and the presence of vesicles in the root cells of some mangroves suggests that fungal symbionts play a role in nutrient uptake; phosphate-solubilizing bacteria associated with the roots and fungi may release phosphate that could be taken up by the fungal hyphae and transferred to the host or taken up directly by the roots.
- The amount of organic detritus potentially available for export depends on net forest primary production, tidal range, the ratio of mangrove to watershed area, lateral trapping, high salinity plugs, total mangrove area, frequency of storms, amount of rainfall, volume of water exchange, and the activities of the litter-feeding fauna. Each system is unique; some mangroves export nutrients and some do not. Where there is a clear pattern of mangrove export, what role this material has as a nutrient subsidy for offshore food webs is unclear. A picture is emerging, however, which indicates that the extent of mangrove influence is restricted to a few kilometers offshore.
- Mass balance estimates for carbon in six mangrove ecosystems show that (1) mangrove ecosystems are net autotrophic, with an average P_g/R ratio of 1.6, (2) gross primary production and net ecosystem production average 383 and 139 mol C m^{-2} $year^{-1}$, (3) exports equate to 1–15% of mangrove net primary production, (4) canopy respiration equates to ≈58% of gross primary production and is probably higher as the data account only for respiration of leaves and does not include stem and root respiration, (5) soil and pelagic respiration is small compared with canopy respiration, and (6) little carbon is buried within the forest floor.
- There is a strong positive correlation between tidal range and net ecosystem production for these six ecosystems implying that tides play an important role in regulating ecosystem production.
- A whole-ecosystem budget for nitrogen exists only for Hinchinbrook Island in northern Australia. The budget shows that the main nitrogen inputs are: (1) nitrogen fixation, (2) tidal exchange, and (3) precipitation. Net tidal exchange is a loss due to the export of litter followed by denitrification and to a lesser extent, sedimentation in the forest and on creek banks. The budget is roughly balanced as nitrogen is greatly conserved within this system.
- Little information is available on cycling of essential elements such as iron, potassium, sodium, magnesium, calcium, zinc, copper, manganese, and vanadium. Some data from Chinese mangroves suggests that the turnover of some elements via litter is faster on average than in most terrestrial forests. The turnover time of an element most likely relates more closely to the rate of primary production than forest age.

- Simulation models and sensitivity analyses for various mangroves suggest that: (1) detritus, nutrients, mangrove biomass are highly sensitive to tidal action, with the most pronounced effect being on the flow of detritus; (2) the strongest influence on mangrove biomass is terrestrial run-off and, secondarily, solar radiation; (3) mangrove production is important with respect to microbial carbon flow rather than within food chain dynamics directly; (4) most energy flow is vested in lower trophic levels; (5) in systems with high planktonic primary production, the main driver of food chains is herbivorous zooplankton. Half of the carbon fixed by phytoplankton is not immediately used by higher trophic levels but shunted into the detritus pool, most of which was directly consumed by fish and other large consumers; (6) lagoonal-mangrove systems are highly productive, but autotrophs other than mangroves play a major energetic role; and (7) in more open, riverine estuaries, mangroves dominate energy flow.
- Models of pristine and reclaimed mangrove forests show that herbivorous and detritivorous pathways are equally important, with human impacts resulting in an increase in the relative importance of algal primary production, herbivorous zooplankton, and meiofauna. Benthic communities within virgin forests are more typically detritus-based, being more dependent on litter. Pristine forests have more energy flow passing through a given area per unit time than reclaimed forests, but a greater percentage of energy flow is lost via respiration in the disturbed forests. Proportionally more energy is imported and exported from a healthy ecosystem with a greater percentage vested in recycling.
- Ecological economic models show that: (1) even a small change in mangrove area will greatly impact on fisheries harvest, (2) the conflicting needs of various users can best we met in the long term if the mangroves are left alone, and (3) exploitation of mangroves for wood and fisheries is currently unsustainable in many areas.
- Comparatively simple models based on empirical data can quantify uppers limits of sustainable use of resources. Sustainable harvesting of mangroves must involve understanding of the level of ecosystem support required to service the growth and maintenance of the harvested stands. In the Matang Reserve in Malaysia, level of ecosystem support is 46.6 ha of total ecosystem area for each ha of forest harvested. At the current harvesting rate, the level of ecosystem support required to sustain each tonne of wood harvested is 2.8 ha. Similar calculations for terrestrial forests (1.0–5.7 ha t^{-1}) suggest that ecological dynamics of mangrove harvesting is similar to that of other forested ecosystems.
- A global model of carbon flow in mangrove ecosystems reveals: (1) the largest flux of carbon is between the trees and atmosphere with a little more than half of gross primary production respired by the trees; (2) ≈40% of net forest primary production is vested in litterfall and wood production with the remaining 60% allocated to root production; (3) total ecosystem respiration is 500 TgC $year^{-1}$, which equates to nearly 70% of forest GPP; (4) carbon burial averages 29 TgC $year^{-1}$ which is about 10% of forest NPP; (5) export of POC and DOC equates to 43 TgC $year^{-1}$ which is about 10–15% of forest NPP; (6) net ecosystem production averages ≈160 TgC $year^{-1}$. However, considering that all of the major inputs

and outputs are already accounted for, this excess carbon may represent DIC export via drainage from soils at ebb tide, (7) if the large production estimate of fine roots (174 TgC $year^{-1}$) is correct, rapid turnover/metabolism of these fine roots would account for much of the excess carbon, (8) mangroves contribute disproportionately to carbon burial in the coastal ocean, as carbon burial in mangroves is greater in proportion to habitat area than salt marshes, seagrass beds, and unvegetated estuarine sediments, and (9) mangroves occupy only about 0.5% of total coastal area globally, but their rates of GPP, NEP, and R account for ≈5–6% of the global total, clearly contributing a small, but disproportionate, share to global carbon cycling.

7.3 Epilogue

Mangrove forests appear to function in many ways more similarly to tropical humid evergreen forests than their supposed coastal counterparts—salt marshes.

These tropical tidal forests are truly unique ecosystems, and it has been a consistent theme of this book to show that mangroves are unique for a variety of structural and functional reasons, not least of which is the overriding fact that they have attributes originating from both terrestrial and marine biomes. Being the only woody plants that grow and thrive in salt water, these trees form true forests ensconced within tropical coastal settings varying from microtidal, stony substrates to megatidal, expansive rich muds that stretch as far as the eye can see. It is sad to note that in these modern times many luxuriant mangrove forests, composed of mature trees that have grown to heights of 20–25 m or more, are now only encountered in pockets of isolated coast.

In my younger days, I remember marvelling at the immense *Bruguiera* and *Rhizophora* forests in the delta of the Fly River in Papua New Guinea, forests so rich in life and vitality that they rivalled many of the region's rainforests in splendor and biological diversity. Standing within one of these forests, I was awed by their sheer size, beauty, and aura of power. Now, most of these tidal forests are gone, felled for wood destined for hearth and home in wealthy nations thirsty for exotic hardwoods.

But all is not lost. While some biodiversity can never be recovered, monospecific forests or stands of initially low diversity have been, and continue to be, successfully replanted. If well tended in their early stages then left alone to develop, some of these reconstituted forests may in time regain much of their lost diversity and magnificence. Regardless of whether or not a mangrove ecosystem is pristine or disturbed by man or nature or both, the functional processes described in this book do and will continue to characterize these important intertidal ecosystems.

References

Aber JD, Melillo JM (2001) Terrestrial ecosystems, 2nd edn. Harcourt, San Diego, CA

Aerts R, Chapin III FS (2000) The mineral nutrition of wild plants revisited: A re-evaluation of processes and patterns. Adv Ecol Res 30:1–67

Aksornkoae S, Panichsuko S, Saraya A, Pinichchat S, Mongkolmol C, Jintana V, Krachivong J, Kongsangchai J, Srisawatt W, Boromthanaratana S, Aguru V, Jintana N, Kooha B, Chungpivat V (1989) Inventory and monitoring on mangroves in Thailand. Final report to the ASEAN-Australia cooperative programme on marine science: Living resources in coastal areas with emphasis on mangrove and coral reef ecosystems, pp 1–112. Bangkok, Thailand

Albright LJ (1976) *In situ* degradation of mangrove tissues. NZ J Mar Freshw Res 10:385–389

Alfaro AC, Thomas F, Sergent L, Duxbury M (2006) Identification of trophic interactions within an estuarine food web (northern New Zealand) using fatty acid biomarkers and stable isotopes. Est Coast Shelf Sci 70:271–286

Allen DE, Dalal RC, Rennenberg H, Meyer RL, Reeves S, Schmidt S (2007) Spatial and temporal variation of nitrous oxide and methane flux between subtropical mangrove sediments and the atmosphere. Soil Biol Biochem 39:622–631

Allen JA, Ewel KC, Keeland BC, Tara T, Smith III TJ (2000) Downed wood in Micronesian mangrove forests. Wetlands 20:169–176

Allison MA, Kepple EB (2001) Modern sediment supply to the lower delta plain of the Ganges-Brahmaputra River in Bangladesh. Geo-Mar Lett 21:66–74

Allison MA, Nittrouer CA, Faria LEC (1995) Rates and mechanisms of shoreface progradation and retreat downdrift of the Amazon River mouth. Mar Geol 125:373–392

Alongi DM (1989) The role of soft-bottom benthic communities in tropical mangrove and coral reef ecosystems. Rev Aquat Sci 1:243–280

Alongi DM (1994) Zonation and seasonality of benthic primary production and community respiration in tropical mangrove forests. Oecologia 98:320–327

Alongi DM (1996) The dynamics of benthic nutrient pools and fluxes in tropical mangrove forests. J Mar Res 54:123–148

Alongi DM (1998) Coastal ecosystem processes. CRC Press, Boca Raton, FL

Alongi DM (2001) The influence of mangrove biomass and production on biogeochemical processes in tropical macrotidal settings. In: Aller JY, Woodin SA, Aller RC (eds) Organism-sediment interactions, pp 223–241. University of South Carolina Press, Columbia

Alongi DM (2002) Present state and future of the world's mangrove forests. Environ Conserv 29: 331–349

Alongi DM (2005a) Mangrove-microbe-soil relations. In: Kristensen E, Haese RR, Kostka JE (eds) Interactions between macro- and microorganisms in marine sediments, pp 85–103. American Geophysical Union, Washington, DC

Alongi DM (2005b) A simple mass balance framework for estimating limits to sustainable mangrove production: Some examples from managed forests in Southeast Asia. Int J Ecol Environ Sci 31:147–155

Alongi DM (2007) The contribution of mangrove ecosystems to global carbon cycling and greenhouse gas emissions. In: Tateda Y (ed) Greenhouse gas and carbon balances in mangrove coastal ecosystems, pp 1–10. Gendai Tosho, Kanagawa, Japan

Alongi DM (2008) Mangrove forests: Resilience, protection from tsunamis, and responses to global climate change. Est Coast Shelf Sci 76:1–13

Alongi DM, de Carvalho NA (2008) The effect of small-scale logging on stand characteristics and soil biogeochemistry in mangrove forests of Timor Leste. For Ecol Manag 255:1359–1366

Alongi DM, Dixon P (2000) Mangrove primary production and above-and below-ground biomass in Sawi Bay, southern Thailand. Phuket Biol Cent Spec Publ 22:31–38

Alongi DM, McKinnon AD (2005) The cycling and fate of terrestrially-derived sediments and nutrients in the coastal zone of the Great Barrier Reef shelf. Mar Pollut Bull 51:239–252

Alongi DM, Sasekumar A (1992) Benthic communities. In: Robertson AI, Alongi DM (eds) Tropical mangrove ecosystems, pp 137–171. American Geophysical Union, Washington, DC

Alongi DM, Boto KG, Robertson AI (1992) Nitrogen and phosphorus cycles. In: Robertson AI, Alongi DM (eds) Tropical mangrove ecosystems, pp 251–292. American Geophysical Union, Washington, DC

Alongi DM, Sasekumar A, Tirendi F, Dixon P (1998) The influence of stand age on benthic decomposition and recycling of organic matter in managed mangrove forests of Malaysia. J Exp Mar Biol Ecol 225:197–218

Alongi DM, Tirendi F, Dixon P, Trott LA, Brunskill GJ (1999) Mineralization of organic matter in intertidal sediments of a tropical semi-enclosed delta. Est Coast Shelf Sci 48:451–467

Alongi DM, Tirendi F, Clough BF (2000a) Below-ground decomposition of organic matter in forests of the mangroves *Rhizophora stylosa* and *Avicennia marina* along the arid coast of Western Australia. Aquat Bot 68:97–122

Alongi DM, Tirendi F, Trott LA, Xuan TT (2000b) Benthic decomposition rates and pathways in plantations of the mangrove *Rhizophora apiculata* in the Mekong delta, Vietnam. Mar Ecol Prog Ser 194:87–101

Alongi DM, Wattayakorn G, Ayukai T, Clough BF, Wolanski E, Brunskill GJ (2000c) An organic carbon budget for mangrove-fringed Sawi Bay, southern Thailand. Phuket Mar Biol Cent Spec Publ 22:79–85

Alongi DM, Wattayakorn G, Pfitzner J, Tirendi F, Zagorskis I, Brunskill GJ, Davidson A, Clough BF (2001) Organic carbon accumulation and metabolic pathways in sediments of mangrove forests in southern Thailand. Mar Geol 179:85–103

Alongi DM, Trott LA, Wattayakorn G, Clough BF (2002) Below-ground nitrogen cycling in relation to net canopy production in mangrove forests of southern Thailand. Mar Biol 140: 855–864

Alongi DM, Clough BF, Dixon P, Tirendi F (2003a) Nutrient partitioning and storage in arid-zone forests of the mangroves *Rhizophora stylosa* and *Avicennia marina*. Trees 17:51–60

Alongi DM, Chong VC, Dixon P, Sasekumar A, Tirendi F (2003b) The influence of fish cage aquaculture on pelagic carbon flow and water chemistry in tidally dominated mangrove estuaries of peninsular Malaysia. Mar Environ Res 55:313–333

Alongi DM, Sasekumar A, Chong VC, Pfitzner J, Trott LA, Tirendi F, Dixon P, Brunskill GJ (2004a) Sediment accumulation and organic material flux in a managed mangrove ecosystem: Estimates of land-ocean-atmosphere exchange in peninsular Malaysia. Mar Geol 208:383–402

Alongi DM, Wattayakorn G, Boyle S, Tirendi F, Payn C, Dixon P (2004b) Influence of roots and climate on mineral and trace element storage and flux in tropical mangrove soils. Biogeochemistry 69:105–123

Alongi DM, Clough BF, Robertson AI (2005a) Nutrient-use efficiency in arid-zone forests of the mangroves *Rhizophora stylosa* and *Avicennia marina*. Aquat Bot 82:121–131

Alongi DM, Pfitzner J, Trott LA, Tirendi F, Dixon P, Klumpp DW (2005b) Rapid sediment accumulation and microbial mineralization in forests of the mangrove *Kandelia candel* in the Jiulongjiang estuary, China. Est Coast Shelf Sci 63:605–618

Alongi DM, Ramanathan AL, Kannan L, Tirendi F, Trott LA, Bala Krishna Prasad M (2005c) Influence of human-induced disturbance on benthic microbial metabolism in the Pichavaram mangroves, Vellar-Coleroon estuarine complex, India. Mar Biol 147:1033–1044

Alongi DM, Trott LA, Rachmansyah, Tirendi F, McKinnon AD, Undu MC (2008) Growth and development of mangrove forests overlying smothered coral reefs, Sulawesi and Sumatra, Indonesia. Mar Ecol Prog Ser 370:97–109

Amarasinghe MD, Balasubramaniam S (1992) Net primary productivity of two mangrove forest stands on the northwestern coast of Sri Lanka. Hydrobiologia 247:37–47

Andersen FØ, Kristensen E (1988) Oxygen microgradients in the rhizosphere of the mangrove *Avicenna marina*. Mar Ecol Prog Ser 44:201–204

Anderson C, Lee SY (1995) Defoliation of the mangrove *Avicennia marina* in Hong Kong: Causes and consequences. Biotropica 27:218–226

Andrade L, Gonzalez AM, Araujo FV, Parahos R (2003) Flow cytometry assessment of bacterioplankton in tropical marine environments. J Microbiol Meth 55:841–850

Andrews TJ, Muller GJ (1985) Photosynthetic gas exchange of the mangrove, *Rhizophora stylosa* Griff in its natural environment. Oecologia 65:449–455

Anthony EJ (2004) Sediment dynamics and morphological stability of estuarine mangrove swamps in Sherbro Bay, West Africa. Mar Geol 20:207–224

Ara K (2001) Daily egg production rate of the planktonic calanoid copepod *Acartia lilljeborgi* Giesbrecht in the Cananéia Lagoon estuarine system, São Paulo, Brazil. Hydrobiologia 445: 205–215

Ara K (2002) Temporal variability and production of *Temora turbinate* (Copepoda: Calanoida) in the Cananéia Lagoon estuarine system, São Paulo, Brazil. Sci Mar 66:399–406

Ashton EC (2002) Mangrove sesarmid crab feeding experiments in Peninsular Malaysia. J Exp Mar Biol Ecol 273:97–119

Atmadja WS, Soerojo (1991) Structure and potential net primary production of mangrove forests at Grajagan and Ujung Kulon, Indonesia. In: Alcala AC (ed) Proceedings of the regional symposium on living resources in coastal areas, pp 441–451. ASEAN-Australia cooperative program in marine sciences, Manila and Townsville

Aubrey DG (1986) Hydrodynamic controls on sediment transport in well-mixed bays and estuaries. In: van de Kreeke J (ed) Physics of shallow estuaries and bays, pp 245–258. Springer, New York

Aucan J, Ridd PV (2000) Tidal asymmetry in creeks surrounded by saltflats and mangroves with small swamp slopes. Wetlands Ecol Manag 8:223–231

Ayukai T, Alongi DM (2000) Pelagic carbon fixation and heterotrophy in shallow coastal waters of Sawi Bay, southern Thailand. Phuket Mar Biol Cent Spec Publ 22:39–50

Ayukai T, Miller D (1998a) Phytoplankton biomass, production and grazing mortality in Exmouth Gulf, a shallow embayment on the arid, tropical coast of Western Australia. J Exp Mar Biol Ecol 225:239–251

Ayukai T, Miller D (1998b) Phytoplankton biomass and production in the Hinchinbrook Channel, northeastern Australia. In: Ayukai T (ed) CO_2 fixation and storage in coastal ecosystems, pp 51–85. Final phase 1 report to the Kansai Electric Power Co. Aust Inst Mar Sci, Townsville

Ayukai T, Miller D, Wolanski E, Spagnol S (1998) Fluxes of nutrients and dissolved and particulate organic carbon in two mangrove creeks in northeastern Australia. Mangr Salt Marsh 2:223–230

Ayukai T, Wolanski E, Wattayakorn G, Alongi DM (2000) Organic carbon and nutrient dynamics in mangrove creeks and adjacent coastal waters of Sawi Bay, southern Thailand. Phuket Mar Biol Cent Spec Publ 22:51–62

Bachok Z, Mfilinge PL, Tsuchiya M (2003) The diet of the mud clam *Geloina coaxans* (Mollusca: Bivalvia) as indicated by fatty acid markers in a subtropical mangrove forest of Okinawa, Japan. J Exp Mar Biol Ecol 292:187–197

Baldocchi DD, Amthor JS (2001) Canopy photosynthesis: History, measurements, and models. In: Roy J, Saugier B, Mooney HA (eds) Terrestrial global productivity, pp 9–31. Academic, New York

Ball MC (1988) Ecophysiology of mangroves. Trees 2:129–142

Ball MC (1996) Comparative ecophysiology of tropical lowland moist rainforest and mangrove forest. In: Mulkey SS, Chazdon RL, Smith AP (eds) Tropical forest plant physiology, pp 461–496. Chapman & Hall, New York

Ball MC, Passioura JB (1995) Carbon gain in relation to water use: Photosynthesis in mangroves. In: Schultze E-D, Caldwell MM (eds) Ecophysiology of photosynthesis, pp 247–259. Springer, Berlin

Ball MC, Chow WS, Anderson JM (1987) Salinity-induced potassium deficiency causes loss of functional photosystem II in leaves of the grey mangrove, *Avicennia marina* through depletion of atrazine-binding polypeptide. Plant Physiol 14:351–361

Ball MC, Cowen JR, Farquhar GD (1988) Maintenance of leaf temperature and the optimisation of carbon gain in relation to water loss in a tropical mangrove forest. Aust J Plant Physiol 15: 263–276

Baltzer F, Allison M, Fromard F (2004) Material exchange between the continental shelf and mangrove-fringed coasts with special reference to the Amazon-Guianas coast. Mar Geol 208: 115–126

Bano N, Nisa M-U, Khan N, Saleem M, Harrison PJ, Ahmed SI, Azam F (1997) Significance of bacteria in the flux of organic matter in the tidal creeks of the mangrove ecosystem of the Indus River delta, Pakistan. Mar Ecol Prog Ser 157:1–12

Barbier EB, Strand I (1998) Valuing mangrove-fishery linkages: A case study of Campeche, Mexico. Environ Res Econ 12:151–166

Barbier EB, Koch EW Silliman BR et al (2008) Coastal ecosystem-based management with non-linear ecological functions and values. Science 319:321–323

Barnes BV, Zak DR, Denton SR, Spurr SH (1998) Forest ecology, 5th edn. Wiley, New York

Barnes J, Ramesh R, Puravja R, Nirmal Rajkumar A, Senthil Kumar B, Krithika K, Ravichandran K, Uher G, Upstill-Goddard RC (2006) Tidal dynamics and rainfall control of N_2O and CH_4 emissions from a pristine mangrove creek. Geophys Res Lett 33:doi 10.1029/2006GL026829

Barnes J, Purvaja R, Ramesh R, Uher G, Upstill-Goddard RC (2007) Nitrous oxide fluxes in Indian mangroves: Tidal production mechanisms, fluxes and global significance. In: Tateda Y (ed) Greenhouse gas and carbon balances in mangrove coastal ecosystems, pp 139–151. Gendai Tosho, Kanagawa, Japan

Barrera-Alba JJ, Gianesella SMF, Moser GAO, Saldanha-Corréa (2008) Bacterial and phytoplankton dynamics in a sub-tropical estuary. Hydrobiologia 598:229–246

Barrett MA, Stiling P (2006) Effects of key deer herbivory on forest communities in the lower Florida Keys. Biol Conserv 129:100–108

Bashan Y, Holguin G (2002) Plant growth-promoting bacteria: A potential tool for arid mangrove reforestation. Trees 16:159–166

Bashan Y, Puente ME, Myrold DD, Tolero G (1998) In vitro transfer of fixed nitrogen from diazotrophic filamentous cyanobacteria to black mangrove seedlings. FEMS Microbiol Ecol 26:165–170

Bauza JF (2007) Emission of greenhouse gases from mangrove forest sediment in Puerto Rico. In: Tateda Y (ed) Greenhouse gas and carbon balances in mangrove coastal ecosystems, pp 165–177. Gendai Tosho, Kanagawa, Japan

Bauza JF, Morell JM, Corredor JE (2002) Biogeochemistry of nitrous oxide production in the red mangrove (*Rhizophora mangle*) forest sediment. Est Coast Shelf Sci 55:697–704

Berger U, Hildenbrandt H (2000) A new approach to spatially explicit modelling of forest dynamics: Spacing, aging, and neighbourhood competition of mangrove trees. Ecol Model 132:287–302

Berger U, Adams M, Grimm V, Hildenbrandt H (2006) Modelling secondary succession of neotropical mangroves: Causes and consequences of growth reduction in pioneer species. Persp Plant Ecol Evol Syst 7:243–252

Bertrand F (1999) Mangrove dynamics in the Rivieres du Sud area, West Africa: An ecogeographic approach. Hydrobiologia 413:115–126

Bingham BL, Young CM (1995) Stochastic events and dynamics of a mangrove root epifaunal community. PSZNI Mar Ecol 16:145–163

Bird ECF (1971) Mangroves as land-builders. Vic Nat 88:189–197

Biswas H, Mukhopadhyay SK, De TK, Sen S, Jana TK (2004) Biogenic controls on the air-water carbon dioxide exchange in the Sunderbans mangrove environment, northeast coast of Bay of Bengal, India. Limnol Oceanogr 49:95–101

Biswas H, Chatterjee A, Mukhopadhya SK, De TK, Sen S, Jana TK (2005) Estimation of ammonia exchange at the land-ocean boundary condition of Sunderbans mangrove, northeast coast of Bay of Bengal, India. Atmos Environ 39:4489–4499

Björkman O, Demmig B, Andrews TJ (1988) Mangrove photosynthesis: Response to high-irradiance stress. Aust J Plant Physiol 15:43–61

Blaber SJM (2002) 'Fish in hot water': The challenges facing fish and fisheries research in tropical estuaries. J Fish Biol 61:1–20

Borges AV (2005) Do we have enough pieces of the jigsaw to integrate CO_2 fluxes in the coastal ocean? Estuaries 28:3–27

Borges AV, Djenidi S, Lacroix G, Théate J, Delille B, Frankignoulle M (2003) Atmospheric CO_2 flux from mangrove surrounding waters. Geophys Res Lett 30:doi:10.1029/2003GL017143

Borges AV, Delille B, Frankignoulle M (2005) Budgeting sinks and sources of CO_2 in the coastal ocean: Diversity of ecosystems counts. Geophys Res Lett 32:L14601, doi:10.1029/2005GL02305

Borges LMS, Cragg SM, Bergot J, Williams JR, Shayler B, Sawyer GS (2008) Laboratory screening of tropical hardwoods for natural resistance to the marine borer *Limnoria quadripunctata*: The role of leachable and non-leachable factors. Holzforsch 62:99–111

Bosire JO, Dahdouh-Guebas F, Kairo JG, Kazungu J, Dehairs F, Koedam N (2005) Litter degradation and C:N dynamics in reforested mangrove plantations at Gazi Bay, Kenya. Biol Conserv 126:287–295

Boto KG (1991) Nutrients and mangroves. In: Connell DW, Hawker DW (eds) Pollution in tropical aquatic systems, pp 129–145. CRC Press, Boca Raton, FL

Boto KG, Bunt JS (1981) Dissolved oxygen and pH relationships in northern Australian mangrove waterways. Limnol Oceanogr 26:1176–1178

Boto KG, Wellington JT (1983) Phosphorus and nitrogen nutritional status of a northern Australian mangrove forest. Mar Ecol Prog Ser 11:63–69

Boto KG, Wellington JT (1988) Seasonal variations in concentrations and fluxes of dissolved organic and inorganic nutrients in a tropical, tidally-dominated, mangrove waterway. Mar Ecol Prog Ser 50:151–160

Boto KG, Saffigna P, Clough BF (1985) Role of nitrate in nitrogen nutrition of the mangrove *Avicennia marina*. Mar Ecol Prog Ser 21:259–265

Boto KG, Alongi DM, Nott ALJ (1989) Dissolved organic carbon-bacteria interactions at sediment-water interface in a tropical mangrove system. Mar Ecol Prog Ser 51:243–251

Bouillon S, Chandra Mohan P, Sreenivas N, Dehairs F (2000) Sources of suspended organic matter and selective feeding by zooplankton in an estuarine mangrove ecosystem as traced by stable isotopes. Mar Ecol Prog Ser 208:79–92

Bouillon S, Koedam N, Raman AV, Dehairs F (2002a) Primary producers sustaining macro-invertebrate communities in intertidal mangrove forests. Oecologia 130:441–448

Bouillon S, Raman AV, Dauby P, Dehairs F (2002b) Carbon and nitrogen stable isotope ratios of subtidal benthic invertebrates in an estuarine mangrove ecosystem (Andhra Pradesh, India). Est Coast Shelf Sci 54:901–913

Bouillon S, Frankignoulle M, Dehairs F, Velimirov B, Eiler A, Abril G, Etcheber H, Borges AV (2003) Inorganic and organic carbon biogeochemistry in the Gautami Godavari estuary (Andhra Pradesh, India) during pre-monsoon: The local impact of extensive mangrove forests. Global Biogeochem Cycle 19:doi:10.1029/2002GB002026

Bouillon S, Koedam N, Baeyens W, Satyanarayana B, Dehairs, F (2004) Selectivity of subtidal benthic invertebrate communities for local microagal production in an estuarine mangrove ecosystem during the post-monsoon period. J Sea Res 51:133–144

Bouillon S, Borges AV, Castañeda-Moya E, Diele K, Dittmat T, Duke NC, Kristensen E, Lee SY, Marchard C, Middelburg JJ, Rivera-Monroy VH, Smith III TJ, Twilley RR (2007a) Mangrove production and carbon sinks: A revision of global budget estimates. Global Biogeochem Cycle 22:GB2013, doi:10.1029/2007GB003OS2

Bouillon S, Dehairs F, Schiettecatte LS, Borges AV (2007b) Biogeochemistry of the Tana estuary and delta (northern Kenya). Limnol Oceanogr 52:46–59

Bouillon S, Middelburg JJ, Dehairs F, Borges AV, Abril G, Flindt MR, Ulomi S, Kristensen E (2007c) Importance of intertidal sediment processes and porewater exchange on the water column biogeochemistry in a pristine mangrove creek (Ras Dege, Tanzania). Biogeosci 4:311–322

Bouillon S, Connolly RM, Lee SY (2008) Organic matter exchange in mangrove ecosystems: Recent insights from stable isotope studies. J Sea Res 59:44–58

Brinkman RM, Massel SR, Ridd PV, Furukawa K (1997) Surface wave attenuation in mangrove forests. In: Pacific coasts and ports'97: Proceedings of the 13th Australasian coastal and ocean engineering conference and the 6th Australasian port and harbour conference, 7–11 September, pp 941–946. Centre for Advanced Engineering, University of Canterbury, Christchurch, New Zealand

Brown AG, Namibar EKS, Cosslter C (1997) Plantations for the tropics—their role, extent, and nature. In: Namibar EKS, Brown AG (eds) Management of soil, nutrients and water in tropical plantation forests, pp 1–24. ACIAR Monograph No. 43. ACIAR, Canberra

Bruenig EF (1990) Tropical forest resources. In: Furtado J, Morgan WB, Pfafflin JR, Ruddle K (eds) Tropical resources: ecology and development, pp 67–96. Harwood Academic, Chur, Switzerland

Brunskill GJ, Zagorskis I, Pfitzner J, Ellison J (2004) Sediment and trace element depositional history from the Ajkwa River estuarine mangroves of Irian Jaya (West Papua), Indonesia. Cont Shelf Res 24:2535–2551

Bryce S, Larcombe P, Ridd PV (1998) The relative importance of landward-directed tidal sediment transport versus freshwater flood events in the Normanby River estuary, Cape York Peninsula, Australia. Mar Geol 149:55–78

Bryce S, Larcombe P, Ridd PV (2003) Hydrodynamic and geomorphological controls on suspended sediment transport in mangrove creek systems, a case study: Cocoa Creek, Townsville, Australia. Est Coast Shelf Sci 56:415–431

Bunt JS (1996) Mangrove zonation: An examination of data from 17 riverine estuaries in tropical Australia. Ann Bot 78:333–341

Bunt JS, Boto KG, Boto G (1979) A survey method of estimating potential levels of mangrove forest primary production. Mar Biol 52:123–128

Burchett MD, Field CD, Pulkownik A (1984) Salinity, growth and root respiration in the grey mangrove, *Avicennia marina*. Physiol Plant 60:113–118

Burford MA, Alongi DM, McKinnon AD, Trott LA (2008) Primary production and nutrients in a tropical macrotidal estuary, Darwin Harbour, Australia. Est Coast Shelf Sci 79:440–448

Burkholder PR, Almodóvar LR (1973) Studies on mangrove algal communities in Puerto Rico. Fla Sci 36:50–66

Burton AJ, Pregistzer KS, Ruess RW, Henrick RL, Allen MF (2002) Root respiration in North American forests: Effects of nitrogen concentration and temperature across biomes. Oecologia 131:559–568

Buskey EJ, Peterson JO, Ambler JW (1996) The swarming behaviour of the copepod *Dioithona oculata*: In situ and laboratory studies. Limnol Oceanogr 44:513–521

Buskey EJ, Deyoe H, Jochem FJ, Villareal TA (2003) Effects of mesozooplankton removal and ammonium addition on planktonic trophic structure during a bloom of the Texas 'brown tide': A mesocosm study. J Plankton Res 25:215–228

Buskey EJ, Hyatt CJ, Speekmann CL (2004) Trophic interactions within the planktonic food web in mangrove channels of Twin Cays, Belize, Central America. Atoll Res Bull 529:1–22

Caffrey JM (2003) Production, respiration and net ecosystem production in U.S. estuaries. Environ Monit Assess 81:207–219

Cahoon DR, Lynch JC (1997) Vertical accretion and shallow subsidence in a mangrove forest of southwestern Florida, U.S.A. Mangr Salt Marsh 1:173–186

Cahoon DR, Hensel P, Rybczyz J, McKee KL, Proffitt CE, Perez BC (2003) Mass tree mortality leads to mangrove peat collapse at Bay Islands, Honduras after Hurricane Mitch. J Ecol 91: 1093–1105

Calbet A (2001) Mesozooplankton grazing effect on primary production: A global comparative analysis in marine ecosystems. Limnol Oceangr 46:1824–1830

Calbet A, Landry MR (2004) Phytoplankton growth, microzooplankton grazing, and carbon cycling in marine systems. Limnol Oceangr 49:51–57

Canfield DE, Kristensen E, Thamdrup B (2005) Aquatic geomicrobiology. Elsevier, Amsterdam

Cannicci S, Burrows D, Fratini S, Smith III TJ, Offenberg J, Dahdouh-Guebas F (2008) Faunal impact on vegetation structure and ecosystem function in mangrove forests: A review. Aquat Bot 89:186–200

Chai PPK (1982) Ecological studies of mangrove forests in Sarawak. PhD dissertation. University of Malaya, Kuala Lumpur

Chambers N, Simmons C, Wackernagel M (2000) Sharing nature's interest: ecological footprints as an indicator of sustainability. Earthscan, London

Chambers RM, Pederson KA (2006) Variation in soil phosphorus, sulfur, and iron pools among south Florida wetlands. Hydrobiologia 569:63–70

Cheeseman JM (1994) Depressions of photosynthesis in mangrove canopies. In: Baker NR, Bowyer JR (eds) Photoinhibition of photosynthesis: from molecular mechanisms to the field, pp 377–389. BIOS Scientific Publishers, Oxford

Cheeseman JM, Lovelock CE (2004) Photosynthetic characteristics of dwarf and fringe *Rhizophora mangle* L in a Belizean mangrove. Plant Cell Environ 7:769–780

Cheeseman JM, Herendeen LB, Cheeseman AT, Clough BF (1997) Photosynthesis and photoprotection in mangroves under field conditions. Plant Cell Environ 20:579–588

Cheng W, Fu S, Susfalk RB, Mitchell RJ (2005) Measuring tree root respiration using ^{13}C natural abundance: Rooting medium matters. New Phytol 167:297–307

Chong VC, Low CB, Ichikawa T (2001) Contribution of mangrove detritus to juvenile prawn nutrition: A duel stable isotope study in a Malaysian mangrove forest. Mar Biol 138:77–86

Christensen B (1978) Biomass and primary production of *Rhizophora apiculata* Bl in a mangrove in southern Thailand. Aquat Bot 4:43–52

Cifuentes LA, Coffin RB, Solorzano L, Cardenas W, Espinoza J, Twilley RR (1996) Isotopic and elemental variations of carbon and nitrogen in a mangrove estuary. Est Coast Shelf Sci 43: 781–800

Clark DA, Brown S, Kicklighter DW, Chambers JQ, Thomlinson JR, Ni J, Holland EA (2001) Net primary production in tropical forests: An evaluation and synthesis of existing field data. Ecol Appl 11:371–384

Clarke PJ, Kerrigan RA (2002) The effects of seed predators on the recruitment of mangroves. J Ecol 90:728–736

Clayton DA (1993) Mudskippers. Oceanogr Mar Biol Annu Rev 31:507–577

Clough BF (1992) Primary productivity and growth of mangrove forests. In: Robertson AI, Alongi DM (eds) Tropical mangrove ecosystems, pp 225–249. American Geophysical Union, Washington, DC

Clough BF (1997) Mangrove ecosystems. In: English S, Wilkinson CR, Baker V (eds) Survey manual for tropical marine resources, 2nd edn. pp 119–196. Aust Inst Mar Sci, Townsville

Clough BF (1998) Mangrove forest productivity and biomass accumulation in Hinchinbrook Channel, Australia. Mangr Salt Marsh 2:191–198

Clough BF, Sim RG (1989) Changes in gas exchange characteristics and water-use efficiency of mangroves in response to salinity and vapour pressure deficit. Oecologia 79:38–44

Clough BF, Boto KG, Attiwill PM (1983) Mangroves and sewage: A re-evaluation. In: Teas HJ (ed) Biology and ecology of mangroves, tasks for vegetation science, vol 8, pp 151–161. Junk, The Hague

Clough BF, Dixon P, Dalhaus O (1997a) Allometric relationships for estimating biomass in multi-stemmed trees. Aust J Bot 45:1023–1031

Clough BF, Ong JE, Gong WK (1997b) Estimating leaf area index and photosynthetic production in canopies of the mangrove *Rhizophora apiculata*. Mar Ecol Prog Ser 159:285–292

Clough BF, Tan DT, Buu DC, Phuong DX, Thao VN (1999) Mangrove forest growth, yield and silvicultural options, appendix XIV. Mixed shrimp farming-mangrove forestry models in the Mekong Delta, Termination report, part B, technical appendices, ACIAR-MOFI project report FIS/94/12, pp 235–251. ACIAR, Canberra

Clough BF, Tan DT, Phuong DX, Buu DC (2000) Canopy leaf area index and litter fall in stands of the mangrove *Rhizophora apiculata* of different age in the Mekong Delta, Vietnam. Aquat Bot 66:311–320

Cocheret de la Morinière E, Nagelkerken I, van der Meij H, van der Velde G (2004) What attracts juvenile coral reef fish to mangroves: Habitat complexity or shade? Mar Biol 144: 139–145

Cole JJ (1999) Aquatic microbiology for ecosystem scientists: New and recycled paradigms in ecological microbiology. Ecosystems 2:215–225

Cooksey KE, Cooksey B (1978) Growth-inhibiting substances in sediment extracts from a subtropical wetland: investigation using a diatom assay. J Phycol 14:347–354

Corredor JE, Morell JM, Bauza J (1999) Atmospheric nitrous oxide fluxes from mangrove sediments. Mar Pollut Bull 38:473–478

Costanza RR, d'Arge R, deGroot R, Farber S, Grasso M, Hannon B (1998) The value of the world's ecosystem services and natural capital. Ecol Econ 25:3–15

Coulter SC, Duarte CM, Tuan MS, Tri NH, Ha HT, Giang LH, Hong PN (2001) Retrospective estimates of net leaf production in *Kandelia candel* mangrove forests. Mar Ecol Prog Ser 221:117–124

Cox EF, Allen JA (1999) Stand structure and productivity of the introduced *Rhizophora mangle* in Hawaii. Estuaries 22:276–284

Cragg SM (1993) Wood break-down in mangrove ecosystems: A review. Papua New Guinea J Agric For Fish 36:30–39

Cruz-Escalona VH, Arreguín-Sánchez F, Zetina-Rejón M (2007) Analysis of the ecosystem structure of Laguna Alvarado, western Gulf of Mexico, by means of a mass balance model. Est Coast Shelf Sci 72:155–167

Curran M (1985) Gas movements in the roots of *Avicennia marina*. (Forsk) Vierh. Aust J Plant Physiol 12:97–108

Dakin WS (1938) The habits and life-history of a penaeid prawn (*Penaeus plebejus*). Proc Zool Soc Lond A 108:163–183

Das AB, Parida A, Basak UC, Das P (2002) Studies on pigments, proteins and photosynthetic rates in some mangroves and mangrove associates from Bhitarkanika, Orissa. Mar Biol 141: 415–422

Davis III SE, Childers DL, Day JW, Rudnick DT, Sklar FH (2001) Wetland-water column exchanges of carbon, nitrogen, and phosphorus in a southern Everglades dwarf mangrove. Estuaries 24: 610–622

Dawes CJ (1996) Macroalgal diversity, standing stock and productivity in a northern mangal on the west coast of Florida. Nova Hedwigia 112:525–535

Dawes CJ, Siar K, Marlett D (1999) Mangrove structure, litter and macroalgal productivity in a northern-most forest of Florida. Mangr Salt Marsh 3:259–267

Day Jr JW, Coronada-Molina C, Vera-Herrera FR, Twilley RR, Rivera-Monroy VH, Alvearez-Guillen H, Day R, Conner W (1996) A 7 year record of above-ground net primary production in a southeastern Mexican mangrove forest. Aquat Bot 55:39–60

de Graaf GJ, Xuan TT (1998) Extensive shrimp farming, mangrove clearance and marine fisheries in the southern provinces of Vietnam. Mang Salt Marsh 2:159–166

Devoe NN, Cole TG (1998) Growth and yield in mangrove forests of the Federated States of Micronesia. For Ecol Manag 103:33–48

Dham VV, Heredia AM, Wafur S, Wafar M (2002) Seasonal variations in uptake and in situ regeneration of nitrogen in mangrove waters. Limnol Oceanogr 47:241–254

Dham VV, Wafar M, Heredia AM (2005) Nitrogen uptake by size-fractionated phytoplankton in mangrove waters. Aquat Microb Ecol 41:281–291

Dittel AI, Epifano CE, Cifuentes LA, Kirchman DL (1997) Carbon and nitrogen sources for shrimp postlarvae fed natural diets from a tropical mangrove system. Est Coast Shelf Sci 45: 629–637

Dittmar T, Lara RJ (2001a) Molecular evidence for lignin degradation in sulfate-reducing mangrove sediments (Amazônia, Brazil). Geochim Cosmochim Acta 65:1417–1428

Dittmar T, Lara RJ (2001b) Do mangroves rather than rivers provide nutrients to coastal environments south of the Amazon River? Evidence from long-term flux measurements. Mar Ecol Prog Ser 213:67–77

Dittmar T, Lara RJ (2001c) Driving forces behind nutrient and organic matter dynamics in a mangrove tidal creek in North Brazil. Est Coast Shelf Sci 52:249–259

Dittmar T, Lara RJ, Kattner G (2001) River or mangrove? Tracing major organic matter sources in tropical Brazilian coastal waters. Mar Chem 73:253–271

Dittmar T, Hertkorn N, Kattner G, Lara RJ (2006) Mangroves, a major source of dissolved organic carbon to the oceans. Global Biogeochem Cycle 20:doi:10.1029/2005GB002570

Doley D, Yates DJ, Unwin GL (1987) Photosynthesis in an Australian rainforest tree, *Argyrodendron peralatum*, during the rapid development and relief of water deficits in dry season. Oecologia 74:441–449

Drechsel P, Zech W (1991) Foliar nutrient levels of broad-leaved tropical trees: A tabular review. Plant Soil 131:29–46

Drechsel P, Zech W (1993) Mineral nutrition of tropical trees. In: Pancel L (ed) Tropical forestry handbook, vol 1, pp 515–567. Springer, Berlin

Drexler JZ, DeCarlo EW (2002) Source water partitioning as a means of characterizing hydrologic function in mangroves. Wetlands Ecol Manag 10:103–113

Dromgoole FI (1988) Carbon dioxide fixation in aerial roots of the New Zealand mangrove *Avicennia marina* var *resinifera*. NZ J Mar Freshw Res 22:617–619

Duarte CM, Thampanya U, Terrados J, Geertz-Hansen O, Fortes MD (1999) The determination of the age and growth of SE Asian mangrove seedlings from internodal counts. Mangr Salt Marsh 3:251–257

Duarte CM, Middelburg JJ, Caraco N (2005) Major role of marine vegetation on the oceanic carbon cycle. Biogeosci 2:1–8

Ducklow HW, Shiah F-K (1993) Bacterial production in estuaries. In: Ford TE (ed) Aquatic microbiology, an ecological approach, pp 261–287. Blackwell Scientific, Boston, MA

Duggan S, McKinnon AD, Carleton JH (2008) Zooplankton in an Australian tropical estuary. Est Coasts 31:455–467

Duke NC (2002) Sustained high levels of foliar herbivory of the mangrove *Rhizophora stylosa* by a moth larva *Doratifera stenosa* (Limacodidae) in northeastern Australia. Wetlands Ecol Manage 10:403–419

Duke NC, Ball MC, Ellison JC (1998) Factors influencing the biodiversity and distributional gradients in mangroves. Global Ecol Biogeogr Lett 7:27–47

Dykyjova D, Ulehlova B (1998) Mineral economy and cycling of minerals in wetlands. In: Westlake DF, Kvet J, Szczepanski A (eds) The production ecology of wetlands, pp 319–366. Cambridge University Press, Cambridge

Eamus D, Myers B, Duff G, Williams D (1999) Seasonal changes in photosynthesis of eight savanna tree species. Tree Physiol 19:665–671

Eisma D (1998) Intertidal deposits: river mouths, tidal flats, and coastal lagoons. CRC Press, Boca Raton, FL

Ellison AM, Farnsworth EJ (1992) The ecology of Belizean mangrove-root fouling communities: Patterns of epibiont distribution and abundance, and effects on root growth. Hydrobiologia 247:87–98

Ellison AM, Farnsworth EJ (2001) Mangrove communities. In: Bertness MD, Gaines SD, Hay ME (eds) Marine community ecology, pp 423–442. Sinauer, Sunderland, MA

Ellison AM, Farnsworth EJ, Twilley RR (1996) Facultative mutualism between red mangroves and root-fouling sponges. Ecology 77:22431–2444

Ellison JC (1993) Mangrove retreat with rising sea-level, Bermuda. Est Coast Shelf Sci 37:75–87

Ellison JC (2005) Holocene palynology and sea-level change in two estuaries in Southern Irian Jaya. Paleogeogr Paleoclimatol Paleoecol 220:291–309

Emmerson WD, McGwynne LE (1992) Feeding and assimilation of mangrove leaves by the crab *Sesarma meinerti* de Man in relation to leaf-litter production in Mgazana, a warm-temperate southern African mangrove swamp. J Exp Mar Biol Ecol 157:41–53

Engel S, Pawlik JR (2005) Interactions among Florida sponges. II. Mangrove habitats. Mar Ecol Prog Ser 303:145–152

Epstein E (1999). Silicon. Annu Rev Plant Physiol Plant Mol Biol 50:641–664

Erickson AA, Saltis M, Bell SS, Dawes CJ (2003) Herbivore feeding preferences as measured by leaf damage and stomatal ingestion: A mangrove crab example. J Exp Mar Biol Ecol 289: 123–138

Fabre A, Fromard F, Trichon V (1999) Fractionation of phosphate in sediments of four representative mangrove stages. Hydrobiologia 392:13–19

Falkowski PG, Fenchel T, Delong EF (2008) The microbial engines that drive Earth's biogeochemical cycles. Science 320:1034–1039

FAO (2003) Status and trends in mangrove area extent worldwide. Forest resources assessment working paper 63. Forest Resources Division, Rome. http://www.fao.org/docrep/007/j1533e/J1533E00.htm

Faraco LFD, da Cunha Lana P (2004) Leaf-consumption levels in subtropical mangroves of Paranaguá Bay (SE Brazil). Wetlands Ecol Manag 12:115–122

Farnsworth EJ (2004) Hormones and shifting ecology throughout plant development. Ecology 85:5–15

Farnsworth EJ, Ellison AM (1991) Patterns of herbivory in Belizean mangrove swamps. Biotropica 23:555–567

Farquhar GD, von Caemmerer S, Berry JA (1980) A biochemical model of photosynthetic CO_2 assimilation in C_3 plants. Planta 149:78–90

Fassbender HW (1998) Long-term studies of soil fertility in cacao-shade trees-agroforestry systems: Results of 15 years of organic matter and nutrient research in Costa Rica. In: Schulte A, Ruhiyat (eds) Soils of tropical forest ecosystems: characteristics, ecology and management, pp 150–158. Springer, Berlin

Faunce CH, Serafy JE (2006) Mangroves as fish habitat: 50 years of field studies. Mar Ecol Prog Ser 318:1–18

Faunce CH, Serafy JE (2008) Growth and secondary production of an eventual reef fish during mangrove residency. Est Coast Shelf Sci 79:93–110

Feller IC (1995) Effects of nutrient enrichment on growth and herbivory of dwarf red mangrove (*Rhizophora mangle* L). Ecol Monogr 65:477–505

Feller IC (2002) The role of herbivory by wood-boring insects in mangrove ecosystems in Belize. Oikos 97:167–176

Feller IC, Mathis WN (1997) Primary herbivory by wood-boring insects along an architectural gradient of *Rhizophora mangle* L. Biotropica 29:440–451

Feller IC, Whigham DF, O'Neill JP, McKee KL (1999) Effects of nutrient enrichment on within-stand cycling in a mangrove forest. Ecology 80:2193–2205

Feller IC, Whigham DF, McKee KM, O'Neill JP (2002) Nitrogen vs phosphorus limitation across an ecotonal gradient in a mangrove forest. Biogeochemistry 62:145–175

Feller IC, Whigham DF, McKee KL, Lovelock CE (2003) Nitrogen limitation of growth and nutrient dynamics in a mangrove forest, Indian River Lagoon, Florida. Oecologia 134:405–414

Feller IC, Lovelock CE, McKee KL (2007) Nutrient addition differentially affects ecological processes of *Avicennia germinans* in nitrogen versus phosphorus limited mangrove ecosystems. Ecosystems 10:347–359

Fenchel T (2008) The microbial loop—25 years later. J Exp Mar Biol Ecol 366:99–103.

Ferreira TO, Otero XL, Vidal-Torrado P, Macias F (2007a) Effects of bioturbation by root and crab activity on iron and sulfur biogeochemistry in mangrove substrate. Geoderma 142:36–46

Ferreira TO, Vidal-Torrado P, Otero XL, Macias F (2007b) Are mangrove forest substrates sediments or soils? A case study in southeastern Brazil. Catena 70:79–91

Finn JT (1976) Measures of ecosystem structure and function derived from analysis of flows. J Theor Biol 56:363–380

Fittkau EJ, Klinge H (1973) On biomass and trophic structure of the central Amazonian rain forest ecosystem. Biotropica 5:2–14

Fleck J, Fitt WK (1999) Degrading mangrove leaves of *Rhizophora mangle* Linne provide a natural cue for settlement and metamorphosis of the upside down jellyfish *Cassiopea xamachana* Bigelow. J Exp Mar Biol Ecol 234:83–94

Folster H, Khanna PK (1997) Dynamics of nutrient supply in plantation soils. In: Sadanandan Nambiar EK, Brown AG (eds) Management of soil, nutrients, and water in tropical plantations. ACIAR Monograph 43, pp 339–378. ACIAR, Canberra

Fratini S, Cannicci S, Vannini M (2000) Competition and interaction between *Neosarmatium smithi* (Crustacea: Grapsidae) and *Terebralia palustris* (Mollusca: Gastropoda) in a Kenyan mangrove. Mar Biol 137:309–316

Friedrichs CT, Lynch DR, Aubrey DG (1992) Velocity asymmetries in frictionally-dominated tidal embayments: Longitudinal and lateral variability. In: Prandle D (ed) Dynamics and exchanges in estuaries and the coastal zone, pp 277–312. American Geophysical Union, Washington, DC

Fromard F, Puig H, Mougin E, Marty G, Betoulle JL, Cadamuro L (1998) Structure, above-ground biomass and dynamics of mangrove ecosystems: New data from French Guiana. Oecologia 115:39–53

Fromard F, Vega C, Priosy C (2004) Half a century of dynamic coastal change affecting mangrove shorelines of French Guiana. A case study based on remote sensing data and field surveys. Mar Geol 208:265–280

Fujimori T (2001) Ecological and silvicultural strategies for sustainable forest management. Elsevier Science, Amsterdam

Furukawa K, Wolanski E (1996) Sedimentation in mangrove forests. Mangr Salt Marsh 1:3–10

Furukawa K, Wolanski E, Mueller H (1997) Currents and sediment transport in mangrove forests. Est Coast Shelf Sci 44:301–310

Ganguly D, Dey M, Mandal SK, De TK, Jana TK (2008) Energy dynamics and its implication to biosphere-atmosphere exchange of CO_2, H_2O and CH_4 in a tropical mangrove forest canopy. Atmos Environ 42:4172–4184

Gattuso J-P, Frankignoulle M, Wollast R (1998) Carbon and carbonate metabolism in coastal aquatic ecosystems. Annu Rev Ecol Syst 29:405–434

Gedney RN, Kapetsky JM, Kuhnhold WW (1982) Training on assessment of coastal aquaculture potential, Malaysia. SCS/GEN/82/35 South China Sea fisheries development and coordination program, Manila

Ghosh S, Jana TK, Singh BN, Choudhury A (1987) Comparative study of carbon dioxide system in virgin and reclaimed mangrove waters of Sunderbans during freshnet. Mahasagar Bull Natl Inst Oceanogr 20:155–161

Giani L, Bashan Y, Holguin G, Strangmann A (1996) Characteristics and methanogenesis of the Balandra lagoon soils, Baja California Sur, Mexico. Geoderma 72:149–160

Gleason SM, Ewel KC (2002) Organic matter dynamics on the forest floor of a Micronesian mangrove forest: An investigation of species composition shifts. Biotropica 34:190–198

Gleason SM, Ewel KC, Hue N (2003) Soil redox conditions and plant-soil relationships in a Micronesian mangrove forest. Est Coast Shelf Sci 56:1065–1074

Gocke K, Hernández C, Giesenhagen H, Hoppe H-G (2004) Seasonal variations of bacterial abundance and biomass and their relation to phytoplankton in the hypertrophic tropical lagoon Ciénaga Grande de Santa Marta, Colombia. J Plankton Res 26:1429–1439

Godhantaraman N (2002) Seasonal variations in species composition, abundance, biomass and estimated production rates of tintinnids at tropical estuarine and mangrove waters, Parangipettai, southeast coast of India. J Mar Syst 36:161–171

Golley FB, Odum HT, Wilson RF (1962) The structure and metabolism of a Puerto Rican red mangrove forest in May. Ecology 43:9–19

Golley FB, McGinnis K, Clements RG, Child GI, Duever MJ (1975) Mineral cycling in a tropical moist forest ecosystem. University of Georgia Press, Athens

Gomes HR, Pant A, Goes JI, Parulekar AH (1991) Heterotrophic utilization of extracellular products of phytoplankton in a tropical estuary. J Plankton Res 13:487–498

Gong WK, Ong JE (1990) Plant biomass and nutrient flux in a managed mangrove forest in Malaysia. Est Coast Shelf Sci 31:516–530

Gong WK, Ong JE, Wong CH, Dhanarajan G (1984) Productivity of mangrove trees and its significance in a managed mangrove ecosystem in Malaysia. In: Soepadmo E, Rao AN,

Macintosh DJ (eds) Proceedings of the UNESCO Asian symposium on mangrove environment: research and management, pp 216–225. University of Malaya, Kuala Lumpur, Malaysia

Gong WK, Ong JE, Wong CH (1991) The light attenuation method for the measurement of potential productivity in mangrove ecosystems In: Alcala A (ed) Proceedings of the ASEAN-Australia regional symposium on living resources in coastal areas, pp 399–406. Manila, Philippines

Gong WK, Ong JE, Clough BF (1992) Photosynthesis in different aged stands of a Malaysian mangrove ecosystem. In: Chou LM, Wilkinson CR (eds) Third ASEAN science and technology week conference proceedings, vol 6, marine science: living coastal resources, 21–23 September, pp 345–351. National University of Singapore and National Science Technology BD, Singapore

Grace J, Malhi Y, Higuchi N, Meir P (2001) Productivty of tropical forests. In: Roy J, Saugier B, Mooney HA (eds) Terrestrial global productivity, pp 401–426. Academic, New York

Grasso M (1998) Ecological-economic model for optimal mangrove trade-off between forestry and fishery production: Comparing a dynamic optimization and a simulation model. Ecol Model 112:131–150

Grubb PJ (1995) Mineral nutrition and soil fertility in tropical rain forests. In: Lugo AE, Lowe C (eds) Tropical forests: management and ecology, pp 308–328. Springer, Berlin

Grubb PJ, Edwards PJ (1982) Studies of mineral cycling in a montone rain forest in New Guinea. III. The distribution of mineral elements in the above-ground material. J Ecol 70:623–648

Guest MA, Connolly RM (2004) Fine-scale movement and assimilation of carbon in saltmarsh and mangrove habitat by resident animals. Aquat Ecol 38:599–609

Guest MA, Connolly RM, Loneragan NR (2004) Carbon movement and assimilation by invertebrates in estuarine habitat at a scale of metres. Mar Ecol Prog Ser 278:27–34

Guest MA, Connolly RM, Lee SY, Loneragan NR, Brietfuss MJ (2006) Carbon movement and assimilation by invertebrates in estuarine habitats: Organic matter transfer not crab movement. Oecologia 148:88–96

Halide H, Brinkman R, Ridd PV (2004) Designing bamboo wave attenuators for mangrove plantations. Indian J Mar Sci 33:220–225

Halle F, Oldeman RAA, Tomlinson PB (1978) Tropical trees and forests-an architectural analysis. Springer, Berlin

Hambrey J (1996) Comparative economics of land use options in mangroves. Aquac Asia 1:10–14

Hamzah L, Hara K, Imamura F (1999) Experimental and numerical study on the effect of mangrove to reduce tsunami. Tohoku J Nat Dis Sci 35:127–132

Harada K, Imamura F (2005) Effects of coastal forest on tsunami hazard mitigation—a preliminary investigation. Adv Nat Tech Haz Res 23:279–292

Harrison PJ, Khan N, Saleem M, Bano N, Nisa M, Ahmed SI, Rizvi N, Azam F (1997) Nutrient and phytoplankton dynamics in two mangrove tidal creeks of the Indus River delta, Pakistan. Mar Ecol Prog Ser 157:13–19

Hauff RD, Ewel KC, Jack J (2006) Tracking human disturbance in mangroves: Estimating harvest rates on a Micronesian island. Wetlands Ecol Manag 14:95–105

Heald EJ (1969) The production of organic detritus in a South Florida estuary. Ph.D. dissertation. University of Miami, FL

Healey MJ, Moll RA, Diallo CO (1988) Abundance and distribution of bacterioplankton in the Gambia River, West Africa. Microb Ecol 16:291–310

Hemminga MA (1995) Interlinkages between eastern African coastal ecosystems. Second semi-annual report on the EC Std-3 Project. Netherlands Institute of Ecology, Yerseke

Hemminga MA, Slim FJ, Kazungu J, Gannsen GM, Nieuwenhuize J, Kruyt NM (1994) Carbon outwelling from a mangrove forest with adjacent seagrass beds and cora lreefs (Gazi Bay, Kenya). Mar Ecol Prog Ser 106:291–301

Hernes PJ, Benner R, Cowie GL, Goñi MA, Bergamaschi BA, Hedges JI (2001) Tannin diagenesis in mangrove leaves from a tropical estuary: A novel molecular approach. Geochim Cosmochim Acta 65:3109–3122

Heron SF, Ridd PV (2001) The use of computational fluid dynamics in predicting the tidal flushing of animal burrows. Est Coast Shelf Sci 52:411–421

Heron SF, Ridd PV (2003) The effect of water density variations on the tidal flushing of animal burrows. Est Coast Shelf Sci 58:137–145

Heron SF, Ridd PV (2008) The tidal flushing of multiple-loop animal burrows. Est Coast Shelf Sci 78:135–144

Hesse PR (1962) Phosphorus fixation in mangrove swamp muds. Nature 193:295–296

Hesse PR (1963) Phosphorus relationships in a mangrove swamp mud with particular reference to aluminium toxicity. Plant Soil 19:205–218

Hessen DO, Anderson TR (2008) Excess carbon in aquatic organisms and ecosystems: Physiological, ecological, and evolutionary implications. Limnol Oceanogr 53:1685–1696

Hicks DB, Burns LA (1975) Mangrove metabolic response to alterations of natural freshwater drainage to southwestern Florida estuaries. In: Walsh GE, Snedaker SC, Teas HJ (eds) Proceedings of the international symposium on biology and management of mangroves, vol 1, pp 238–255. Institute of Food and Agricultural Science, University of Florida, Gainesville, FL

Hiraishi T, Harada K (2003). Greenbelt tsunami prevention in South Pacific region. Rep Port Airport Res Inst 42:1–23

Hiremath AJ, Ewel JJ, Cole TG (2002) Nutrient-use efficiency in three fast-growing tropical trees. For Sci 48:662–672

Holguin G, Bashan Y (1996) Nitrogen-fixation by *Azospirillum brasilense* is promoted when co-cultured with a mangrove rhizosphere bacterium (*Staphylococcus* sp.). Soil Biol Biochem 28:1651–1660

Holguin G, Vazquez P, Bashan Y (2001) The role of sediment microorganisms in the productivity, conservation, and rehabilitation of mangrove ecosystems: An overview. Biol Fertil Soils 33: 265–278

Holmboe N, Kristensen E, Andersen FØ (2001) Anoxic decomposition in sediments from a tropical mangrove forest and the temperate Wadden Sea: Implications of N and P addition experiments. Est Coast Shelf Sci 53:125–140

Holmer M, Kristensen E, Banta G, Hansen K, Jensen MH, Bussarawit N (1994) Biogeochemical cycling of sulfur and iron in sediments of the south-east Asian mangrove, Phuket Island, Thailand. Biogeochemistry 25:1–17

Holmer M, Andersen FØ, Holmboe N, Kristensen E, Thongtham N (1999) Transformation in the Bangrong mangrove forest-seagrass bed system, Thailand: Seasonal and spatial variations in benthic metabolism and sulfur biogeochemistry. Aq Micro Ecol 20:203–212

Holmer M, Andersen FØ, Kristensen E, Thongtham N (2001) Transformation and exchange processes in the Bangrong mangrove forest-seagrass bed system, Thailand: Seasonal variations in benthic primary production and nutrient dynamics. Wetlands Ecol Manage 9:141–158

Hopkinson Jr CS, Smith EM (2005) Estuarine respiration: An overview of benthic, pelagic, and whole system respiration. In: del Giorgio P, Williams, PJ le B (eds) Respiration in aquatic ecosystems, pp 122–146. Oxford University Press, Oxford

Hossain M, Othman S, Bujang JS, Kusnan M (2008) Net primary productivity of *Bruguiera parviflora* (Wight & Arn.) dominated mangrove forest at Kuala Selangor, Malaysia. For Ecol Manag 255:179–182

Howarth RR, Marino R, Lane J, Cole JJ (1988) Nitrogen fixation in freshwater, estuarine, and marine ecosystems. 1. Rates and importance. Limnol Oceanogr 33:669–687

Hsieh H-L, Chen C-P, Chen Y-G, Yang H-H (2002) Diversity of benthic organic matter flows through polychaetes and crabs in a mangrove estuary: $\delta^{13}C$ and $\delta^{34}S$ signals. Mar Ecol Prog Ser 227:145–155

Huitric M, Folke C, Kautsky N (2002) Development and government policies of the shrimp farming industry in Thailand in relation to mangrove ecosystems. Environ Res Econ 40:441–455

Hwang YH, Chen SC (2001) Effects of ammonium, phosphate, and salinity on growth, gas exchange characteristics, and ionic contents of seedlings of the mangrove *Kandelia candel* (L) Druce. Bot Bull Acad Sinica 42:131–139

Iizumi H (1986) Soil nutrient dynamics. In: Cragg S, Polunin N (eds) Workshop on mangrove ecosystem dynamics. UNDP/UNESCO regional project (RAS/79/002), pp 171–180. UNESCO, New Delhi

Imamura M, Tateda Y, Ishii T (2007) Emission of nitrous oxide from coastal water and sediments of mangrove forest ecosystems. In: Tateda Y (ed) Greenhouse gas and carbon balances in mangrove coastal ecosystems, pp 179–188. Gendai Tosho, Kanagawa, Japan

Ishida A, Toma T, Marjenah (1999a) Leaf gas exchange and chlorophyll fluorescence in relation to leaf angle, azimuth, and canopy position in the tropical pioneer tree, *Macaranga conifera*. Tree Physiol 19:117–124

Ishida A, Toma T, Marjenah (1999b) Limitation of leaf carbon gain by stomatal and photochemical processes in the top canopy of *Macaranga conifera*, a tropical pioneer tree. Tree Physiol 19:467–473

Ishimatsu A, Hishida Y, Takita T, Kanda T, Oikawa S, Takeda T, Huat KK (1998) Mudskippers store air in their burrows. Nature 391:237–238

Islam MdS, Haque M (2004) The mangrove-based coastal and nearshore fisheries of Bangladesh: Ecology, exploitation and management. Rev Fish Biol Fish 14:153–180

Janssen R, Padilla JE (1999) Preservation or conversion? Valuation and evaluation of a mangrove forest in the Philippines. Environ Res Econ 14:297–331

Jayasekera R (1991) Chemical composition of the mangrove, *Rhizophora mangle* L. J Plant Physiol 138:119–121

Jennerjahn TC, Ittekkot V (1997) Organic matter in sediments in the mangrove areas and adjacent continental margins of Brazil: I. Amino acids and hexosamines. Oceanol Acta 20:359–369

Jennerjahn TC, Ittekkot V (2002) Relevance of mangroves for the production and deposition of organic matter along tropical continental margins. Naturwissenschaften 89:23–30

Jiménez JA, Lugo AE, Cintron G (1985) Tree mortality in mangrove forests. Biotropica 17: 177–185

Jordan CF (1985) Nutrient cycling in tropical forest ecosystems. Wiley, Chicester

Joshi GV, Sontakke S, Bhosale I, Waghmode AP (1984) Photosynthesis and photorespiration in mangroves. In: Teas HJ (ed) Physiology and management of mangroves, pp 210–238. Dr. W. Junk, The Hague

Joye SB, Lee RY (2004) Benthic microbial mats: Important sources of fixed nitrogen and carbon to the Twin Cays, Belize ecosystem. Atoll Res Bull 53:1–24

Jyothibabu R, Madhu NV, Jayalakshmi KV, Balachandran KK, Shiyas CA, Martin GD, Nair KKC (2006) Impact of freshwater influx on microzooplankton mediated food web in a tropical estuary (Cochin backwaters—India). Est Coast Shelf Sci 69:505–518

Kathiresan K (2000) A review of studies on Pichavarum mangroves, southeast India. Hydrobiologia 430:185–205

Kathiresan K, Bingham BL (2001) Biology of mangroves and mangrove ecosystems. Adv Mar Biol 40:81–251

Kathiresan K, Rajendran N (2002) Fishery resources and economic gain in three mangrove areas on the south-east coast of India. Fish Manage Ecol 9:277–283

Kay JJ, Graham LA, Ulanowicz RE (1989) A detailed guide to network analysis. In: Wulff F, Fields JG, Mann KH (eds) Network analysis in marine ecology: methods and applications, pp 15–61. Springer, Berlin

Kenzo T, Ichie T, Yoneda R, Kitahashi Y, Watanabe Y, Ninomiya I, Koike T (2004) Interspecific variation of photosynthesis and leaf characteristics in canopy trees of five species of Dipterocarpaceae in a tropical rain forest. Tree Physiol 24:1187–1192

Kibirige I, Perissinotto R, Nozais C (2002) Alternative food sources of zooplankton in a temporarily-open estuary: Evidence from $\delta^{13}C$ and $\delta^{15}N$. J Plankton Res 24:1089–1095

Kieckbusch DK, Koch MS, Serafy JE, Anderson WT (2004) Trophic linkages among primary producers and consumers in fringing mangroves of subtropical lagoons. Bull Mar Sci 74: 271–285

Kimmerer WJ, McKinnon AD (1987) Zooplankton in a marine bay. II. Vertical migration to maintain horizontal distributions. Mar Ecol Prog Ser 41:53–60

Kimmins JP (2004) Forest ecology, 3rd edn. Prentice-Hall, Upper Saddle River, NJ.

Kirita H, Hozumi K (1973) Estimation of the total chlorophyll amount and its seasonal change in a warm-temperate evergreen oak forest at Minimata, Japan. Jpn J Ecol 23:195–200

Kitheka JU (1996) Water circulation and coastal trapping of brackish water in a tropical mangrove-dominated bay in Kenya. Limnol Oceanogr 41:169–176

Kitheka JU (1998) Groundwater outflow and its linkage to coastal circulation in a mangrove-fringed creek in Kenya. Est Coast Shelf Sci 47:63–75

Kitheka JU, Ohowa BO, Mwashote BM, Shimbira WS, Mwaluma JM, Kazungu JM (1996) Water circulation dynamics, water column nutrients and plankton productivity in a well-flushed tropical bay in Kenya. J Sea Res 35:257–268

Kitheka JU, Mwashote BM, Ohowa BO, Kamau J (1999) Water circulation, groundwater outflow and nutrient dynamics in Mida Creek, Kenya. Mangr Salt Marsh 3:135–146

Kitheka JU, Ongwenyi GS, Mavuti KM (2003) Fluxes and exchange of suspended sediment in tidal inlets draining a degraded mangrove forest in Kenya. Est Coast Shelf Sci 56:655–667

Kitaya Y, Yabuki K, Kiyota M, Tani A, Hirano T, Aiga I (2002) Gas exchange and oxygen concentration in pneumatophores and prop roots of four mangrove species. Trees 16:155–158

Koch BP, Harder J, Lara RJ, Kattner G (2005) The effect of selective microbial degradation on the composition of mangrove derived pentacyclic triterpenols in surface sediments. Org Geochem 36:273–285

Koch MS, Snedaker SC (1997) Factors influencing *Rhizophora mangle* L seedling development in Everglades carbonate soils. Aquat Bot 59:87–98

Koch MS, Benz RE, Rudnick DT (2001) Solid-phase phosphorus pools in highly organic carbonate sediments of Northeastern Florida Bay. Est Coast Shelf Sci 52:279–291

Koch V, Wolff M (2002) Energy budget and ecological role of mangrove epibenthos in the Caeté estuary, North Brazil. Mar Ecol Prog Ser 228:119–130

Komiyama A, Ogino K, Sabhasri S (1987) Root biomass of mangrove forests in southern Thailand. 1. Estimation by the trench method and the zonal structure of root biomass. J Trop Ecol 3:97–108

Komiyama A, Ong JE, Poungparn S (2008) Allometry, biomass, and productivity of mangrove forests: A review. Aquat Bot 89:128–137

Koné YJ-M, Borges AV (2008) Dissolved inorganic carbon dynamics in the waters surrounding forested mangroves of the Ca Mau Province (Vietnam). Est Coast Shelf Sci 77:409–421

Königer M, Harris GC, Virgo A, Winter K (1995) Xanthophyll-cycle pigments and photosynthetic capacity in tropical forest species: A comparative field study on canopy, gap and understory plants. Oecologia 104:270–290

Kothamasi D, Kothamasi S, Bhattacharayya A, Chander Kuhad R, Babu CR (2006) Arbusicular mycorrhizae and phosphate-solubilizing bacteria of the rhizophore of the mangrove ecosystem of Great Nicobar Island, India. Biol Fertil Soils 42:358–361

Krause GH, Virgo A, Winter K (1995) Photoinhibition of photosynthesis in plants growing in natural tropical forest gaps. A chlorophyll fluorescence study. Planta 197:583–591

Krause TEC, Dahlgren RA, Zasoki RJ (2003) Tannins in nutrient dynamics of forest ecosystems: A review. Plant Soil 256:41–66

Krauss KW, Doyle TW, Twilley RR, Smith III TJ, Whelen KRT, Sullivan JK (2005) Woody debris in the mangrove forests of South Florida. Biotropica 37:9–15

Krauss KW, Doyle TW, Twilley RR, Rivera-Monroy VH, Sullivan JK (2006) Evaluating the relative contributions of hydroperiod and soil fertility on growth of south Florida mangroves. Hydrobiologia 569:311–324

Krauss KW, Keeland B, Allen J, Ewel K, Johnson D (2007a) Effects of season, rainfall, and hydrogeomorphic setting on mangrove tree growth in Micronesia. Biotropica 39:161–170

Krauss KW, Young PJ, Chambers JL, Doyle TW, Twilley RR (2007b) Sap flow characteristics of neotropical mangroves in flooded and drained soils. Tree Physiol 27:775–783

Krauss KW, Lovelock CE, McKee KL, López-Hoffman L, Ewe SHL, Sousa WP (2008) Environmental drivers in mangrove establishment and early development: A review. Aquat Bot 89:105–127

Kreuzwieser J, Buchholz J, Rennenberg H (2003) Emission of methane and nitrous oxide by Australian mangrove ecosystems. Plant Biol 5:423–431

Kristensen E (2007) Carbon balance in mangrove sediments: The driving processes and their controls. In: Tateda Y (ed) Greenhouse gas and carbon balances in mangrove coastal ecosystems, pp 61–78. Gendai Tosho, Kanagawa, Japan

Kristensen E (2008) Mangrove crabs as ecosystem engineers; with emphasis on sediment processes. J Sea Res 59:30–43

Kristensen E, Alongi DM (2006) Control by fiddler crabs (*Uca vocans*) and plant roots (*Avicennia marina*) on carbon, iron, and sulfur biogeochemistry in mangrove sediment. Limnol Oceanogr 51:1557–1571

Kristensen E, Pilgaard R (2001) The role of fecal pellet deposition by leaf-eating sesarmid crabs on litter decomposition in a mangrove sediment (Phuket, Thailand). In: Aller JY, Woodin SA, Aller RC (eds) Organism-sediment relations, pp 369–384. University of South Carolina Press, Columbia

Kristensen E, Suraswadi P (2002) Carbon, nitrogen and phosphorus dynamics in creek water of a southeast Asian mangrove forest. Hydrobiologia 474:197–211

Kristensen E, Andersen FØ, Kofoed LH (1988) Preliminary assessment of benthic community metabolism in a south-east Asian mangrove swamp. Mar Ecol Prog Ser 48:137–145

Kristensen E, Devol AH, Ahmed SI, Saleem M (1992) Preliminary study of benthic metabolism and sulfate reduction in a mangrove swamp of the Indus Delta, Pakistan. Mar Ecol Prog Ser 90:287–297

Kristensen E, King GM, Holmer M, Banta GT, Jensen MH, Hansen K, Bussarawit (1994) Sulfate reduction, acetate turnover and carbon metabolism in sediments of the Ao Nam Bor mangrove, Phuket, Thailand. Mar Ecol Prog Ser 199:245–255

Kristensen E, Holmer M, Banta GT, Jensen MH, Hansen K (1995) Carbon, nitrogen and sulfur cycling in sediments of the Ao Nam Bor mangrove forest, Phuket, Thailand: A review. Phuket Mar Biol Cent Res Bull 60:37–64

Kristensen E, Jensen MH, Banta GT, Hansen K, Holmer M, King GM (1998) Transformation and transport of inorganic nitrogen in sediments of a southeast Asian mangrove forest. Aquat Micro Ecol 15:165–175

Kristensen E, Andersen FØ, Holmboe N, Holmer M, Thongtham N (2000) Carbon and nitrogen mineralization in sediments of the Bangrong mangrove area, Phuket, Thailand. Aquat Micro Ecol 22:199–213

Kristensen E, Bouillon S, Dittmar T, Marchand C (2008) Organic carbon dynamics in mangrove ecosystems: A review. Aquat Bot 89:201–219

Kruitwagen G, Nagelkerkan I, Lugendo BR, Pratap HB, Wendelaar Bonga SE (2007) Influence of morphology and amphibious life-style on the feeding ecology of the mudskipper *Periophthalmus argentilineatus*. J Fish Biol 71:39–52

Krumme U, Liang T-H (2004) Tidal-induced changes in a copepod-dominated zooplankton community in a macrotidal mangrove channel in northern Brazil. Zool Stud 43:404–414

Kurosawa K, Suzuki Y, Tateda Y, Sugito S (2003) A model of the cycling and export of nitrogen in Fukido mangrove in Ishigaki Island. J Chem Eng Jpn 36:411–416

Kutner MB (1975) Seasonal variation and phytoplankton distribution in Cananeia region. In: Walsh GE, Snedaker SC, Teas HJ (eds) Proceedings international symposium on the biology and management of mangroves, vol 1, pp 153–169. University of Florida, Gainsville, FL

Lacerda LD (1992) Carbon burial in mangrove sediments, a potential source of carbon to the sea during events of sea-level change. In: Lacerda LD, Turcq B, Knoppers B, Kjerfve B (eds) Paleoclimatic changes and the carbon cycle, pp 107–114. Soc Bras Geoquim, Rio de Janeiro, Brazil

Lacerda LD, Ittekkot V, Patchineelam SR (1995) Biogeochemistry of mangrove soil organic matter: A comparison between *Rhizophora* and *Avicennia* soils in South-eastern Brazil. Est Coast Shelf Sci 40:713–720

Lacerda LD, Conde JE, Kjerfve B, Alvarez-León R, Alarcón C, Polanía J (2002) American mangroves. In: Lacerda LD (ed) Mangrove ecosystems: function and management, pp 1–62. Springer, Berlin

Lai DY, Lam KC (2008) Phosphorus retention and release by sediments in the eutrophic Mai Po Marshes, Hong Kong. Mar Pollut Bull 57:349–356

Lampitt RS, Gamble JC (1982) Diet and respiration of the small planktonic marine copepod *Oithona nana*. Mar Biol 62:185–190

Landry M (2001) Microbial loops. In: Steele JH, Thorpe S, Turekian K (eds) Encyclopedia of ocean sciences, pp 1763–1770. Academic, London

Larcombe P, Ridd PV (1996) Dry season hydrodynamics and sediment transport in mangrove creeks. In: Pattariatchi C (ed) Mixing processes in estuaries and coastal seas, pp 388–404. American Geophysical Union, Washington, DC

Larsson J, Folke C, Kautsky N (1994) Ecological limitations and appropriation of ecosystem support by shrimp farming in Colombia. Environ Manag 18:663–676

Larsson R, Padilla JE (1999) Preservation or conversion? Valuation and evaluation of a mangrove forest in the Philippines. Environ Res Econ 14: 297–331

Latief H, Hadi S (2007) The role of forests and trees in protecting coastal areas against tsunamis. In: Braatz S, Fortuna S, Broadhead J, Leslie R (eds) Coastal protection in the aftermath of the Indian Ocean tsunami: what role for forests and trees?, pp 5–35. Proceedings of the Regional Technical Workshop, Khao Lak, Thailand

Leakey ADB, Press MC, Scholes JD (2003) High-temperature inhibition of photosynthesis is greater under sunflecks than uniform irradiance in a tropical rain forest tree seedling. Plant Cell Environ 26:1681–1690

Lee C-W, Bong C-W (2006) Carbon flux through bacteria in a eutrophic tropical environment: Port Klang waters. In: Wolanski E (ed) The environment of Asia Pacific harbours, pp 329–345. Springer, Dordrecht

Lee C-W, Bong C-W (2007) Bacterial respiration, growth efficiency and protist grazing rates in mangrove waters in Cape Rachado, Malaysia. Asian J Water Environ Pollut 4:11–16

Lee RY, Joye SB (2006) Seasonal patterns of nitrogen fixation and denitrification in oceanic mangrove habitats. Mar Ecol Prog Ser 307:127–141

Lee SY (1989) The importance of sesarminae crabs *Chiromanthes* spp and inundation frequency on mangrove (*Kandelia candel* (L) Druce) leaf litter turnover in a Hong Kong tidal shrimp pond. J Exp Mar Biol Ecol 131:23–43

Lee SY (1990) Primary productivity and particulate organic matter flow in an estuarine mangrove-wetland in Hong Kong. Mar Biol 106:453–463

Lee SY (1995) Mangrove outwelling—a review. Hydrobiologia 295:203–212

Lee SY (1997) Potential trophic importance of the faecal material of the mangrove sesarmine crab *Sesarma messa*. Mar Ecol Prog Ser 159:275–284

Lee SY (1998) Ecological role of grapsid crabs in mangrove ecosystems: A review. Mar Freshw Res 49:335–343

Lee SY (2004) Relationship between mangrove abundance and tropical prawn production: A re-evaluation. Mar Biol 145:943–949

Lee SY (2008) Mangrove macrobenthos: Assemblages, services, and linkages. J Sea Res 59:16–29

Lee YS, Kaur B, Broom MJ (1984) The effect of palm oil effluent on the nutrient status and planktonic primary production of a Malaysian mangrove inlet. In: Soepadmo E, Rao AN, Macintosh DJ (eds) Proceedings of the UNESCO Asian symposium on mangrove environment: research and management, pp 575–591. University of Malaya, Kuala Lumpur, Malaysia

Legendre L, Rivkin RB (2008) Planktonic food webs: Microbial hub approach. Mar Ecol Prog Ser 365:289–309

Leh CMU, Sasekumar A (1984) Feeding ecology of prawns in shallow waters adjoining mangrove shores. In: Soepadmo E, Rao AN, Macintosh DJ (eds) Proceedings of the Asian symposium on mangrove environment: research and management, pp 331–353. Ardyas, University of Malaya and UNESCO

Leh CMU, Sasekumar A (1985) The food of sesarmine crabs in Malaysian mangrove forests. Malay Nat J 39:135–145

Lekphet S, Nitisoravut S, Adsvakulchai S (2005) Estimating methane emissions from mangrove area in Ranong Province, Thailand. Songklanakarin J Sci Technol 27:153–163

Li MS (1997) Nutrient dynamics of a Futian mangrove forest in Shenzhen, south China. Est Coast Shelf Sci 45:463–472

Lin G, Sternberg L daS (1992) Comparative study of water uptake and photosynthetic gas exchange between scrub and fringe red mangroves, *Rhizophora mangle* L. Oecologia 90:399–403

Lin H-J, Shao K-T, Kuo S-R, Hsieh H-L, Wong S-L, Chen I-M, Lo W-T, Hung J-J (1999) A trophic model of a sandy barrier lagoon at Chiku in southwestern Taiwan. Est Coast Shelf Sci 48:575–588

Lin P (1999) Mangrove ecosystems in China. Science Press, Beijing

Lindquist ES, Carroll CR (2004) Differential seed and seedling predation by crabs: Impacts on tropical coastal forest composition. Oecologia 141:661–671

Litton CM, Raich JW, Ryan MG (2007) Carbon allocation in forest ecosystems. Global Change Biol 13:1–21

Loneragan NR, Ahmad Adnan N, Connolly RM, Manson FJ (2005) Prawn landings and their relationship with the extent of mangroves and shallow waters in western peninsular Malaysia. Est Coast Shelf Sci 63:187–200

Lopez OR, Kursar TA (1999) Flood tolerance of four tropical tree species. Tree Physiol 19: 925–932

López-Hoffman L, Anten NPR, Martinez-Ramos M, Ackerly DD (2007) Salinity and light interactively affect neotropical mangrove seedlings at the leaf and whole plant levels. Oecologia 150:545–556

Lovelock CE (2008) Soil respiration and belowground carbon allocation in mangrove forests. Ecosystems 11:342–354

Lovelock CE, Ball MC (2002) Influence of salinity on photosynthesis of halophytes. In: Läuchli A, Lüttge U (eds) Salinity: environment-plants-molecules, pp 315–339. Kluwer, Utrecht

Lovelock CE, Feller IC (2003) Photosynthetic performance and resource utilization of two mangrove species coexisting in hypersaline scrub forest. Oecologia 134:455–462

Lovelock CE, Virgo A, Popp M, Winter K (1999) Effects of elevated CO_2 concentrations on photosynthesis, growth and reproduction of branches of the tropical canopy tree species, *Luehea seemannii* Tr. & Planch. Plant Cell Environ 22:49–59

Lovelock CE, Ball MC, Feller IC, Engelbrecht BMJ, Ewe ML (2006a) Variation in hydraulic conductivity of mangroves: Influence of species, salinity, and nitrogen and phosphorus availability. Physiol Plantarum 127:457–464

Lovelock CE, Feller IC, Ball MC, Engelbrecht BMJ, Ewe ML (2006b) Differences in plant function in phosphorus- and nitrogen-limited mangrove ecosystems. New Phytol 172:514–522

Lovelock CE, Ruess RW, Feller IC (2006c) Fine root respiration in the mangrove *Rhizophora mangle* over variation in forest structure and nutrient availability. Tree Physiol 26:1601–1606

Lovelock CE, Feller IC, Ball MC, Ellis J, Sorrell B (2007) Testing the growth rate vs geochemical hypothesis for latitudinal variation in plant nutrients. Ecol Lett 10:1154–1163

Lu C, Wong Y, Tam NFY, Lovelock CE, Feller IC, McKee KL, Engelbrecht BMJ, Ball MC (2004) The effect of nutrient enrichment on growth, photosynthesis and hydraulic conductance of dwarf mangroves in Panama. Funct Ecol 18:25–33

Lu Y, Cui S, Lin P (1998) Preliminary studies on methane fluxes in Hainan mangrove communities. Chin J Oceanol Limnol 16:57–64

Lugendo BR, Nagelkerken I, van der Velde G, Mgaya YD (2006) The importance of mangroves, mud and sand flats, and seagrass beds as feeding areas for juvenile fishes in Chwaka Bay, Zanzibar: Gut content and stable isotope analyses. J Fish Biol 69:1639–1661

Lugo AE, Evink G, Brinson MM, Broce A, Snedaker SC (1975) Diurnal rates of photosynthesis, respiration, and transpiration in mangrove forests of South Florida. In: Golley FB, Medina E (eds) Tropical ecological systems: trends in terrestrial and aquatic research, pp 335–350. Springer, Berlin

Lugo AE, Sell M, Snedaker SC (1976) Mangrove ecosystem analysis. In: Patten BC (ed) Systems analysis and simulation in ecology, vol IV, pp 113–145. Academic, New York

Lugo AE, Brown S, Brinson MM (1988) Forested wetlands in freshwater and salt-water environments. Limnol Oceanogr 33:894–909

Lüttge U (1997) Physiological ecology of tropical plants. Springer, Berlin
Luyssaert S, Inglima I, Jung M et al. (2007) CO_2 balance of boreal, temperate, and tropical forests derived from a global database. Global Change Biol 13:2509–2537
Lyimo TJ, Pol A, Op den Camp HJM (2002) Methane emission, sulphide concentration and redox potential profiles in Mtoni mangrove sediment, Tanzania. West Ind Ocean J Mar Sci 1:71–80
Lynch JC, Meriwether JR, McKee BA, Vera-Herrera F, Twilley RR (1989) Recent accretion in mangrove ecosystems based on ^{137}Cs and ^{210}Pb. Estuaries 12:284–299
Machado W, Gueiros BB, Lisboa-Filho SD, Lacerda LD (2005) Trace metals in mangrove seedlings: Role of iron plaque formation. Wetlands Ecol Manag 13:199–206
Machiwa JF (2000) δ^{13}C signatures of flora, macrofauna and sediment of a mangrove forest partly affected by sewage wastes. Tanz J Sci 26:16–28
Machiwa JF, Hallberg RO (2002) An empirical model of the fate of organic material in a mangrove forest partly affected by anthropogenic activity. Ecol Model 147:69–83
Macintosh DJ (1977) Quantitative sampling and production estimates of fiddler crabs in a Malaysian mangrove. Mar Res Indones 18:59–73
Macintosh DJ (1984) Ecology and productivity of Malaysian mangrove crab populations (Decapoda: Brachyura). In: Soepadmo E, Rao AN, Macintosh DJ (eds) Proc Asian symp on mangrove environment: research and management, pp 315–341. Ardyas, University of Malaya and UNESCO
Maie N, Pisani O, Jaffé R (2008) Mangrove tannins in aquatic ecosystems: Their fate and possible influence on dissolved organic carbon and nitrogen cycling. Limnol Oceanogr 53:160–171
Manickchand-Heileman S, Arreguín-Sánchez F, Lara-Domínguez A, Soto LA (1998) Energy flow and network analysis of Terminos Lagoon, SW Gulf of Mexico. J Fish Biol 53 (Suppl A): 179–197
Manson FJ, Loneragan NR, Skilleter GA, Phinn SR (2005a) An evaluation of the evidence for linkages between mangroves and fisheries: A synthesis of the literature and identification of research directions. Oceanogr Mar Biol Annu Rev 43:483–513
Manson FJ, Loneragan NR, Harch BD, Skilleter GA, Williams L (2005b) A broad-scale analysis of links between coastal fisheries production and mangrove extent: A case-study for northeastern Australia. Fish Res 74:69–85
Manzoli S, Jackson RB, Trofymow JA, Porporato A (2008) The global stoichiometry of litter nitrogen mineralization. Science 321:684–686
Marchand C, Lallier-Vergès E, Baltzer F (2003) The composition of sedimentary organic matter in relation to the dynamic features of a mangrove-fringed coast in French Guiana. Est Coast Shelf Sci 56:119–130
Marchand C, Disnar JR, Lallier-Vergès E, Lottier N (2005) Early diagenesis of carbohydrates and lignin in mangrove sediments subject to variable redox conditions (French Guiana). Geochim Cosmochim Acta 69:131–142
Marenco RA, Gonçalves J F deC, Vieira G (2001) Leaf gas exchange and carbohydrates in tropical trees differing in successional status in two light environments in central Amazonia. Tree Physiol 21:1311–1318
Maria GL, Sridhar KR (2004) Fungal colonization of immersed wood in mangroves of the southwest coast of India. Can J Bot 82:1409–1418
Marten GG, Polovina JJ (1982) A comparative study of fish yields from various tropical ecosystems. In: Pauly D, Murphy GI (eds) Theory and management of tropical fisheries, pp 255–285. ICLARM, Manila
Martin CE, Loeschen VS (1993) Photosynthesis in the mangrove species *Rhizophora mangle* L: No evidence for CAM cycling. Photosynthetica 28:391–400
Martosubroto P, Naamin N (1977) Relationship between tidal forests (mangroves) and commercial shrimp production in Indonesia. Mar Res Indonesia 18:81–86
Massel SR, Furukawa K, Brinkman RM (1999) Surface wave propagation in mangrove forests. Fluid Dyn Res 24:219–249
Matsui N (1998) Estimated stocks of organic carbon in mangrove roots and sediments in Hinchinbrook Channel, Australia. Mangr Salt Marsh 2:199–204

Mazda Y, Ikeda Y (2006) Behavior of the groundwater in a riverine-type mangrove forest. Wetlands Ecol Manage 14:477–488

Mazda Y, Yokochi H, Sato Y (1990) Groundwater flow in the Bashita-Minato mangrove area, and its influence on water and bottom water properties. Est Coast Shelf Sci 31:621–638

Mazda Y, Kanazawa N, Wolanski E (1995) Tidal asymmetry in mangrove creeks. Hydrobiologia 295:51–58

Mazda Y, Magi M, Kogo M, Hong PH (1997a) Mangroves as a coastal protection from waves in the Tong King delta, Vietnam. Mangr Salt Marsh 1:127–135

Mazda Y, Wolanski E, King B, Sase A, Ohtsuka D, Magi M (1997b) Drag force due to vegetation in mangrove swamps. Mangr Salt Marsh 1:193–199

Mazda Y, Kanazawa N, Kurokawa T (1999) Dependence of dispersion on vegetation density in a tidal creek-mangrove swamp system. Mangr Salt Marsh 3:59–66

Mazda Y, Kobashi D, Okada S (2005) Tidal-scale hydrodynamics within mangrove swamps. Wetlands Ecol Manag 13:647–655

Mazda Y, Magi M, Ikeda Y, Kurokawa T, Asano T (2006) Wave reduction in a mangrove forest dominated by *Sonneratia* sp. Wetlands Ecol Manag 14:365–378

Mazda Y, Wolanski E, Ridd PV (2007) The role of physical processes in mangrove environments: Manual for the preservation and utilization of mangrove ecosystems. TERRAPUB, Tokyo

McGuiness KA (1997) Seed predation in a tropical mangrove forest: A test of the dominance-predation model in northern Australia. J Trop Ecol 13:293–302

McKee KL (1993) Soil physicochemical patterns and mangrove species distribution – reciprocal effects? J Ecol 81:477–487

McKee KL (1996) Growth and physiological responses of neotropical mangrove seedlings to root zone hypoxia. Tree Physiol 16:883–889

McKee KL (2001) Root proliferation in decaying roots and old root channels: A nutrient conservation mechanism in oligotrophic mangrove forests? J Ecol 89:876–887

McKee KL, Faulkner PL (2000) Restoration of biogeochemical function in mangrove forests. Restoration Ecol 8:247–259

McKee KL, Mendelssohn IA (1987) Root metabolism in the black mangrove (*Avicennia germinans* (L): Response to hypoxia. Environ Exp Bot 27:147–156

McKee KL, Mendelssohn IA, Hester MW (1988) Reexamination of pore water sulfide concentrations and redox potentials near the aerial roots of *Rhizophora mangle* and *Avicennia germinans*. Am J Bot 75:1352–1359

McKee KL, Feller IC, Popp M, Wanek W (2002) Mangrove isotopic ($\delta^{15}N$ and $\delta^{13}C$) fractionation across a nitrogen vs phosphorus limitation gradient. Ecology 83:1065–1075

McKee KL, Cahoon DR, Feller IC (2007) Caribbean mangroves adjust to rising sea level through biotic controls on change in soil elevation. Global Ecol Biogeogr 16:545–556

McKillup SC, McKillup RV (1997) An outbreak of the moth *Achaea serva* (Fabr) on the mangrove *Excoecaria agallocha* (L). Pan-Pacific Entomol 73:184–185

McKinnon AD, Ayukai T (1996) Copepod egg production and food resources in Exmouth Gulf, Western Australia. Mar Freshw Res 47:595–603

McKinnon AD, Klumpp DW (1998) Mangrove zooplankton of North Queensland, Australia. II. Copepod egg production and diet. Hydrobiologia 362:145–160

McKinnon AD, Trott LA, Alongi DM, Davidson A (2002a) Water column production and nutrient characteristics in mangrove creeks receiving shrimp farm effluent. Aquac Res 33:55–73

McKinnon AD, Trott LA, Cappo M, Miller DK, Duggan S, Speare P, Davidson A (2002b) The trophic fate of shrimp farm effluent in mangrove creeks of North Queensland, Australia. Est Coast Shelf Sci 55:655–671

McKinnon AD, Smit N, Townsend S, Duggan S (2006) Darwin Harbour: Water quality and ecosystem structure in a tropical harbour in the early stages of development. In: Wolanski E (ed) The environment of Asia Pacific harbours, pp 433–459. Springer, Dordrecht

Medina E, Cuevas E (1989) Patterns of nutrient accumulation and release in Amazonian forests of the upper Rio Negro basin. In: Proctor J (ed) Mineral nutrients in tropical forest and savanna ecosystems, pp 217–240. Blackwell Science, Oxford

Mehlig U (2001). Aspects of tree primary production in an equatorial mangrove forest in Brazil. ZMT Contribution No 14. Cent Trop Mar Ecol, Bremen

Meyer RL, Risgaard-Petersen N, Allen DE (2005) Correlation between anammox activity and microscale distribution of nitrite in a subtropical mangrove sediment. Appl Environ Microbiol 71:6142–6149

Meyer RL, Allen DE, Schmidt S (2008) Nitrification and denitrification as sources of sediment nitrous oxide production: A microsensor approach. Mar Chem 110:68–76

Meziane T, Tsuchiya M (2000) Fatty acids as tracers of organic matter in the sediment and food web of a mangrove/intertidal flat ecosystem, Okinawa, Japan. Mar Ecol Prog Ser 200: 49–57

Meziane T, Sanabe MC, Tsuchiya M (2002) Role of fiddler crabs of a subtropical intertidal flat on the fate of sedimentary fatty acids. J Exp Mar Biol Ecol 270:191–201

Mfilinge PL, Meziane T, Bachok Z, Tsuchiya M (2003) Fatty acids in decomposing mangrove leaves: Microbial activity, decay and nutritional quality. Mar Ecol Prog Ser 265:97–105

Mfilinge PL, Meziane T, Bachok Z, Tsuchiya M (2005) Litter dynamics and particulate organic matter outwelling from a subtropical mangroves in Okinawa Island, South Japan. Est Coast Shelf Sci 63:301–313

Middelburg JJ, Nieuwenhuize J, Slim FJ, Ohowa B (1996) Sediment biogeochemistry in an East African mangrove forest (Gazi Bay, Kenya). Biogeochemistry 34:133–155

Middelburg JJ, Duarte CM, Gattuso J-P (2005) Respiration in coastal benthic communities. In: del Giorgio PA, leB Williams, PJ (eds) Respiration in aquatic ecosystems, pp 206–224. Oxford University Press, Oxford

Middleton BA, McKee KL (2001) Degradation of mangrove tissues and implications for peat formation in Belizean island forests. J Ecol 89:818–828

Miller PC (1972) Bioclimate, leaf temperature, and primary production in red mangrove canopies in south Florida. Ecology 53:22–45

Miller PC (1975) Simulation of water relations and net photosynthesis in mangroves of southern Florida.In: Snedaker SC, Teas HJ (eds) Proceedings of the international symposium on biology and management of mangroves, vol 2, pp 615–631. University of Florida, Gainesville, FL

Millero FJ, Hiscock WT, Huang F, Roche M, Zhang JZ (2001) Seasonal variation of the carbonate system in Florida Bay. Bull Mar Sci 68:101–123

Moberg F, Rönnback P (2003) Ecosystem services of the tropical seascape: Interactions, substitutions and restoration. Ocean Coast Manag 46:27–46

Mohammed SM, Johnstone RW (1995) Spatial and temporal variations in water column nutrient concentrations in a tidally dominated mangrove creek: Chwaka Bay, Zanzibar. Ambio 24: 482–486

Mokhtari M, Savari A, Rezai H, Kochanian P, Bitaab A (2008) Population ecology of the fiddler crab, *Uca lactea annulipes* (Decapoda: Ocypodidae) in Sirik mangrove estuary, Iran. Est Coast Shelf Sci 76:273–281

Monji N (2007) Characteristics of CO_2 flux over a mangrove forest. In: Tateda Y (ed) Greenhouse gas and carbon balances in mangrove coastal ecosystems, pp 197–208. Gendai Tosho, Kanagawa, Japan

Monji N, Hamotani K, Hirano T, Tabuki K, Jintana V (1996) Characteristics of CO_2 flux over a mangrove forest of southern Thailand in rainy season. J Agric Meteorol 52:149–154

Monji, N, Hamotani K, Tosa R, Fukagawa T, Yabuki K, Hirano T, Jintana V, Piriyayota S, Nishimiya A, Iwasaki M (2002a) CO_2 and water vapour flux evaluations by modified methods over a mangrove forest. J Agric Meteorol 58:63–68

Monji N, Hamotani K, Hamada Y, Agata Y, Hirano T, Yabuki K, Jintana V, Piriyayota S, Nishimiya A, Iwasaki M (2002b) Exchange of CO_2 and heat between mangrove forest and the atmosphere in wet and dry seasons in southern Thailand. J Agric Meteorol 58:71–77

Morell JM, Corredor JE (1993) Sediment nitrogen trapping in a mangrove lagoon. Est Coast Shelf Sci 37:203–212

Muda DA, Mustafa NMSN (2003) A working plan for the Matang mangrove forest reserve, Perak. State Forestry Dept Perak, Ipoh, Malaysia

Mukhopadhyay SK, Biswas H, De TK, Jana TK (2006) Fluxes of nutrients from the tropical River Hooghly at the land-ocean boundary of Sunderbans, NE coast of Bay of Bengal, India. J Mar Syst 62:9–21

Muñoz-Hincapié M, Morell JM, Corredor JE (2002) Increase of nitrous oxide flux to the atmosphere upon nitrogen addition to red mangrove sediments. Mar Pollut Bull 44:992–996

Mumby PJ, Edwards AJ, Arlas-González J, Lindeman KC, Blackwell PG, Galt A, Gorczynska MI, Harborne AR, Pescod CL, Renken H, Wabnitz CCC, Llewellyn G (2004) Mangroves enhance the biomass of coral reef fish communities in the Caribbean. Nature 427:533–536

Murach D, Ruhiyat D, Iskandar E, Schulte A (1998) Fine root inventories in dipterocarp forests and plantations in East Kalimantan, Indonesia. In: Schulte A, Ruhiyat D (eds) Soils of tropical forest ecosystems, pp 186–191. Springer, Berlin

Muzuka ANN, Shunula JP (2006) Stable isotope compositions of organic carbon and nitrogen of two mangrove stands along the Tanzanian coastal zone. Est Coast Shelf Sci 66:447–458

Mwashote BM, Jumba IO (2002) Quantitative aspects of inorganic nutrient fluxes in the Gazi Bay (Kenya): Implications for coastal ecosystems. Mar Pollut Bull 44:1194–1205

Naamin N (1990) The ecological and economic roles of Segara Anakan, Indonesia, as a nursery ground of shrimp. In: Chou LM, Khoo T-E et al. (eds) Towards an integrated management of tropical coastal resources, pp 119–130. ICLARM, Manila

Nagelkerken I, Blaber SJM, Bouillon S, Green P, Haywood M, Kirton LG, Meynecke J-O, Pawlik J, Penrose HM, Sasekumar A, Somerfield PJ (2008) The habitat function of mangroves for terrestrial and marine fauna. Aquat Bot 89:155–185

Naidoo G (1987) Effects of salinity and nitrogen on growth and water relations in the mangrove, *Avicennia marina* (Forsk) Vierh. New Phytol 107:317–325

Naidoo G (1990) Effects of nitrate, ammonium and salinity on growth of the mangrove *Bruguiera gymnorrhiza* (L) Lam. Aquat Bot 38:209–219

Naidoo G, Rogalla H, von Willert DJ (1998) Field measurements of gas exchange in *Avicennia marina* and *Bruguiera gymnorrhiza*. Mangr Salt Marsh 2:99–107

Naidoo G, Tuffers AV, von Willert DJ (2002) Changes in gas exchange and chlorophyll fluorescence characteristics of two mangroves and a mangrove associate in response to salinity in the natural environment. Trees 16:140–146

Naidoo G, Steinke TD, Mann FD, Bhatt A, Gairola S (2008) Epiphytic organisms on the pneumatophores of the mangrove *Avicennia marina*: Occurrence and possible function. Afr J Plant Sci 2:12–15

Nambiar EKS, Brown AG (eds) (1997) Management of soil, nutrients and water in tropical plantation forests. ACIAR Monograph No 43, ACIAR, Canberra

Nayar S (2006) Spatio-temporal fluxes in particulate carbon in a tropical coastal lagoon. Environ Monit Assess 112:53–68

Naylor E (2006) Orientation and navigation in coastal and estuarine zooplankton. Mar Freshw Behav Physiol 39:13–24

Nedwell DB, Blackburn TH, Wiebe WJ (1994) Dynamic nature of the turnover of organic carbon, nitrogen and sulphur in the sediments of a Jamaican mangrove forest. Mar Ecol Prog Ser 110:223–231

Newell RIE, Marshall N, Sasekumar A, Chong VC (1995) Relative importance of benthic microalgae, phytoplankton, and mangroves as sources of nutrition for penaeid prawns and other coastal invertebrates from Malaysia. Mar Biol 123:595–606

Newell SY (1996) Established and potential impacts of eukaryotic mycelial decomposers in marine/terrestrial ecotones. J Exp Mar Biol Ecol 200:187–206

Nichol CJ, Rascher U, Matsubara S, Osmond B (2006) Assessing photosynthetic efficiency in an experimental mangrove canopy using remote sensing and chlorophyll fluorescence. Trees 20:9–15

Nickerson DJ (1999) Trade-offs of mangrove area development in the Philippines. Ecol Econ 28:279–298

Nielsen MG, Christian K, Henriksen PG, Birkmose D (2006) Respiration by mangrove ants *Camponotus anderseni* during nest submersion associated with tidal inundation in northern Australia. Physiol Entomol 31:120–126

Nielsen T, Andersen FØ (2003) Phosphorus dynamics during decomposition of mangrove (*Rhizophora apiculata*) leaves in sediments. J Exp Mar Biol Ecol 293:73–88

Nixon SW (1988) Physical energy inputs and the comparative ecology of lake and marine ecosystems. Limnol Oceanogr 33:1005–1025

Nordhaus I, Wolff M, Diele K (2006) Litter processing and population food intake of the mangrove crab *Ucides cordatus* in a high intertidal forest in northern Brazil. Est Coast Shelf Sci 67:239–250

Nunes, RA, Simpson JH (1985) Axial convergence in a well-mixed estuary. Est Coast Shelf Sci 20: 637–649

Nwoboshi LC (1984) Growth and nutrient requirements in a teak plantation age series in Nigeria. II. Nutrient accumulation and minimum annual requirements. For Sci 30:223–239

Nygren P (1995) Leaf CO_2 exchange of *Erythrina poeppigiana* (Leguminosae: Phaseolae) in humid tropical field conditions. Tree Physiol 15:71–83

Nyland RD (2001) Silviculture: concepts and applications. McGraw-Hill, New York

Odum EP (1968) A research challenge: Evaluating the productivity of coastal and estuarine water. In: Proceedings of the second sea grant congress, pp 63–64. University of Rhode Island, Graduate School of Oceanography, Kingston

Odum EP (1969) A strategy for ecosystem development. Science 164:262–269

Odum EP (2000) Tidal marshes as outwelling/pulsing systems. In: Weinstein MP, Kreeger DA (eds) Concepts and controversies in tidal marsh ecology, pp 3–7. Kluwer, Dordrecht

Odum EP, Barrett GW (2005) Fundamentals of ecology, 5th edn. Brooks/Cole, Belmont, CA

Odum HT (1973) Energy, ecology, economics. Ambio 2:220–227

Odum HT (1983) Systems ecology. Wiley-Interscience, New York

Odum WE, Heald EJ (1972) Trophic analyses of an estuarine mangrove community. Bull Mar Sci 22:671–737

Odum WE, Heald EJ (1975) The detritus-based food web of an estuarine mangrove community. In: Cronin LE (ed) Estuarine research, pp 265–286. Academic, New York

Odum WE, Fisher JS, Pickrel JC (1979) Factors controlling the flux of particulate organic carbon from estuarine wetlands. In: Livingston RJ (ed) Ecological processes in coastal and marine systems, pp 69–79. Plenum Press, New York

Odum WE, McIvor CC, Smith III TJ (1982) The ecology of mangroves of South Florida: a community profile. US Fish and Wildlife Service, Office of Biological Services, Washington DC, FWS/OBS-81/24

Odum WE, Odum EP, Odum HT (1995) Nature's pulsing paradigm. Estuaries 18:547–555

Offenberg J, Macintosh DJ, Nielsen MG (2006) Indirect ant protection against crab herbivory: Damage-induced susceptibility to crab grazing may lead to its reduction on ant-colonized trees. Func Ecol 20:52–57

Olafsson E, Buchmayer S, Skov MW (2002) The East African decapod crab *Neosarmatium meinerti* (de Man) sweeps mangrove floors clean of leaf litter. Ambio 31:569–573

Ong JE, Gong WK, Wong CH, Dhanarajan G (1984) Contributions of aquatic productivity in a managed mangrove ecosystem in Malaysia. In: Soepadmo E, Rao AN, Macintosh DJ (eds) Proceedings of the Asian symposium on mangrove environment: research and management, pp 209–215. Ardyas, University of Malaya and UNESCO

Ong JE, Gong WK, Wong CH (1985) Seven years of productivity studies in a Malaysian managed mangrove forest: Then what? In: Bardsley KN, Davie JDS, Woodruffe CD (eds) Coasts and tidal wetlands of the Australian monsoon region, pp 213–223. Australian National University Press, Darwin

Ong JE, Gong WK, Clough BF (1995) Structure and productivity of a 20-year-old stand of *Rhizophora apiculata* Bl mangrove forest. J Biogeogr 22:417–424

Ortiz M, Wolff M (2004) Qualitative modelling for the Caeté estuary (North Brazil): A preliminary approach to an integrated eco-social analysis. Est Coast Shelf Sci 61:243–250

Ovalle ARC, Rezende CE, Lacerda LD, Silva CAR (1990) Factors affecting the hydrochemistry of a mangrove tidal creek, Sepetiba Bay, Brazil. Est Coast Shelf Sci 31:639–650

Ovalle ARC, Rezende CE, Carvalho CEV, Jennerjahn TC, Ittekkot V (1999) Biogeochemical characteristics of coastal waters adjacent to small river-mangrove systems, Brazil. Geo-Mar Lett 19:179–185

Ozaki K, Takashima S, Suko O (2000) Ant predation suppresses populations of the scale insect *Aulacaspismarina* in natural mangrove forests. Biotropica 32:764–768
Pandey JS, Khanna P (1998) Sensitivity analysis of a mangrove ecosystem model. J Environ Syst 26:57–72
Parida AK, Das AB, Mittra B (2004) Effects of salt on growth, ion accumulation, photosynthesis and leaf anatomy of the mangrove, *Bruguiera parviflora*. Trees 18:167–174
Parkinson RW, DeLaune RD, White JR (1994) Holocene sea-level rise and the fate of mangrove forests within the wider Caribbean region. J Coast Res 10:1077–1086
Passioura JB, Ball MC, Knight JH (1992) Mangroves may salinise the soil and in so doing limit their transpiration rate. Funct Ecol 6:476–481
Patanaponpaiboon P, Poungparn S (2000) Photosynthesis of mangroves at Ao Sawi- Thung Kha, Chumphon, Thailand. Phuket Mar Biol Cent Spec Publ 22:27–30
Pauly D, Ingles J (1986) The relationship between shrimp yields and intertidal (mangrove) areas: A reassessment. In: IOC/ FAO workshop on recruitment in tropcial coastal demersal communities, pp 227–284. IOC, UNESCO, Paris
Paw JN, Chua T-E (1991) An assessment of the ecological and economic impact of mangrove conversion in Southeast asia. In: Chou LM (ed) Towards an integrated management of tropical coastal resources, pp 201–212. ICLARM, Manila
Peña EJ, Zingmark R, Nietch C (1999) Comparative photosynthesis of two species of intertidal epiphytic macroalgae on mangrove roots during submersion and emersion. J Phycol 35: 1206–1214
Pérez-España H, Galván-Magaña F, Abitia-Cárdenas LA (1998) Growth, consumption, and productivity of the California killifish in Ojo de Liebre Lagoon, Mexico. J Fish Biol 52: 1068–1077
Perry DA, Oren R, Hart SC (2008) Forest ecosystems, 2nd edn. Johns Hopkins University Press, Baltimore, MD
Phillis YA, Andriantiatsaholiniana LA (2001) Sustainability: An ill-defined concept and its assessment using fuzzy logic. Ecol Econ 29:235–252
Piou C, Berger U, Hildenbrandt H, Feller IC (2008) Testing the intermediate disturbance hypothesis in species-poor systems: A simulation experiment for mangrove forests. J Veget Sci 19: 417–424
Poovachiranon S, Tantichodok P (1991) The role of sesarmid crabs in the mineralisation of leaf litter of *Rhizophora apiculata* in a mangrove, southern Thailand. Phuket Mar Biol Cent Res Bull 56:63–74
Popp M (1995) Salt resistance in herbaceous halophytes and mangroves. Prog Bot 56:416–429
Popp M, Polania J, Weiper M (1993) Physiological adaptations to different salinity levels in mangroves. In: Leith H, Al Masoom, A (eds) Towards the rational use of high salinity tolerant plants, vol 1, pp 217–224. Kluwer, Utrecht
Poret N, Twilley RR, Rivera-Monroy VH, Coronado-Molina C (2007) Belowground decomposition of mangrove roots in Florida Coastal Everglades. Est Coasts 30:491–496
Potts M (1979) Nitrogen-fixation (acetylene reduction) associated with communities of heterocystous and non-heterocystous blue-green algae in the mangrove forests of Sinai. Oecologia 39:359–374
Potts M, Whitton BA (1980) Vegetation of the intertidal zone of the lagoon of Aldabra, with particular reference to the photosynthetic prokaryotic communities. Proc R Soc Lond B 208: 13–33
Pradeep Ram AS, Nair S, Chandramohan D (2003) Seasonal shift in net ecosystem production in a tropical estuary. Limnol Oceanogr 48:1601–1607
Procheş Ş, Marshall DJ (2002) Epiphytic algal cover and sediment deposition as determinants of arthropod distribution and abundance on mangrove pneumatophores. J Mar Biol Ass UK 82: 937–942
Procheş Ş, Marshall DJ, Ugrasen K, Ramcharan A (2001) Mangrove pneumatophore arthropod assemblages and temporal patterns. J Mar Biol Ass UK 81:545–552
Proctor J (1987) Nutrient cycling in primary and old secondary rain forests. Appl Geogr 7:135–152

Proctor J (ed) (1989) Mineral nutrients in tropical forest and savanna ecosystems. Blackwell Science, Oxford

Proffitt CE, Devlin DJ (2005) Grazing by the intertidal gastropod *Melampus coffeus* greatly increases mangrove leaf litter degradation rates. Mar Ecol Prog Ser 296:209–218

Proisy C, Mougin E, Fromard F, Karam MA (2000) Interpretation of polarimetric signatures of mangrove forest. Remote Sens Environ 71:56–66

Proisy C, Mougin E, Fromard F, Karam MA, Trichon V (2002) On the influence of canopy structure on the polarimetric radar response from mangrove forest. Int J Remote Sens 23: 4197–4210

Purvaja R, Ramesh R (2000) Natural and anthropogenic effects on phytoplankton primary productivity in mangroves. Chem Ecol 17:41–58

Purvaja R, Ramesh R (2001) Natural and anthropogenic methane emission from coastal wetlands of south India. Environ Manag 27:547–557

Purvaja R, Ramesh R, Frenzel P (2004) Plant-mediated methane emission from an Indian mangrove. Global Change Biol 10:1825–1834

Putland JN, Iverson RL (2007) Microzooplankton: Major herbivores in an estuarine plankton food web. Mar Ecol Prog Ser 345:63–73

Putz FE, Chan HT (1986) Tree growth, dynamics, and productivity in a mature mangrove forest in Malaysia. For Ecol Manag 17:211–230

Quartel S, Kroon A, Augustinus PGEF, Van Santen PV, Tri NH (2007) Wave attenuation in coastal mangroves in the Red River Delta, Vietnam. J Asian Earth Sci 29:576–584

Raison LK, Brown AG, Flinn D (2001) Criteria and indicators for sustainable forest management. Oxford University Press, Oxford

Ralison OH, Borges AV, Dehairs F, Middelburg JJ, Bouillon S (2008) Carbon biogeochemistry of the Betsiboka estuary (north-western Madagascar). Org Geochem 39:137–149

Ram ASP, Nair S, Chandrmohan D (2003) Seasonal shift in net ecosystem production in a tropical estuary. Limnol Oceanogr 48:1601–1607

Ramesh R, Purvaja R, Neetha V, Divia J, Barnes J, Upstill-Goddard RC (2007) CO_2 and CH_4 emissions from Indian mangroves and its surrounding waters. In: Tateda Y (ed) Greenhouse gas and carbon balances in mangrove coastal ecosystems, pp 153–164. Gendai Tosho, Kanagawa, Japan

Ratanasermpong S, Disbunchhong D, Charuppat T, Ongsomwang S (2000) Assessment of mangrove ecosystem change using remote sensing and GIS technology in Ao Sawi-Thung Kha, Chumphon, Thailand. Phuket Mar Biol Cent Spec Publ 22:3–14

Ravikumar S, Kathiresan K, Thadedus Maria Ignatiammal S, Babu MS, Shanthy S (2004) Nitrogen-fixing azotobacters from mangrove habitat and their utility as marine biofertilizers. J Exp Mar Biol Ecol 312:5–17

Ray S (2008) Comparative study of virgin and reclaimed islands of Sunderban mangrove ecosystem through network analysis. Ecol Model 215:207–216

Ray S, Ulanowicz RE, Majee NC, Roy AB (2000) Network analysis of a benthic food web model of a partly reclaimed island in the Sunderbans mangrove ecosystem, India. J Biol Syst 8:263–278

Reich PB, Walters MB, Ellsworth DS (1997) From tropics to tundra: Global convergence in plant functioning. Proc Nat Acad Sci USA 94:13730–13734

Reich PB, Walters MB, Tjoelker MG, Vanderklein D, Buschena C (1998) Photosynthesis and respiration rates depend on leaf and root morphology in nine boreal tree species differing in relative growth rate. Funct Biol 12:395–405

Ricard M (1984) Primary production in mangrove lagoon waters. In: Por FD, Dor I (eds) Hydrobiology of the mangal, pp 163–178. Dr. W. Junk, The Hague

Richey JE, Melack JM, Aufdenkampe AK, Ballester VM, Hess LL (2002) Outgassing from Amazonian rivers and wetlands as a large tropical source of atmospheric CO_2. Nature 416: 617–620

Ridd PV (1996) Flow through animal burrows in mangrove creeks. Est Coast Shelf Sci 43: 617–625

Ridd PV, Sam R (1996) Profiling groundwater salt concentrations in mangrove swamps and tropical salt flats. Est Coast Shelf Sci 43:627–635

Ridd PV, Stieglitz T (2002) Dry season salinity changes in arid estuaries fringed by mangroves and saltflats. Est Coast Shelf Sci 54:1039–1049

Ridd PV, Wolanski E, Mazda Y (1990) Longitudinal diffusion in mangrove-fringed tidal creeks. Est Coast Shelf Sci 31:541–554

Ridd PV, Stieglitz T, Larcombe P (1998) Density-driven secondary circulation in a tropical mangrove estuary. Est Coast Shelf Sci 47:621–632

Rivera-Monroy VH, Twilley RR (1996) The relative role of denitrification and immobilization in the fate of inorganic nitrogen in mangrove sediments (Terminos Lagoon, Mexico). Limnol Oceangr 41:284–296

Rivera-Monroy VH, Day JW, Twilley RR, Vera-Herrera F, Coronado-Molina C (1995a) Flux of nitrogen and sediment in a fringe mangrove forest in Terminos Lagoon, Mexico. Est Coast Shelf Sci 40:139–160

Rivera-Monroy VH, Twilley RR, Boustany RG, Day JW, Vera-Herrera F, del Carmen Ramirez M (1995b) Direct denitrification in mangrove sediments in Terminos Lagoon, Mexico. Mar Ecol Prog Ser 126:97–109

Rivera-Monroy VH, Madden CJ, Day JW, Twilley RR, Vera-Herrera F, Alvarez-Guillén (1998) Seasonal coupling of a tropical mangrove forest and an estuarine water column/enhancement of aquatic primary productivity. Hydrobiologia 379:41–53

Robertson AI (1986) Leaf-burying crabs: Their influence on energy flow and export from mixed mangrove forests (*Rhizophora* spp.) in northeastern Australia. J Exp Mar Biol Ecol 102:237–248

Robertson AI, Alongi DM (1995) Role of riverine mangrove forests in organic carbon export to the coastal tropical ocean: A preliminary mass balance for the Fly delta (Papua New Guinea). Geo-Mar Lett 15:134–139

Robertson AI, Blaber SJM (1992) Plankton, epibenthos and fish communities. In: Robertson AI, Alongi DM (eds) Tropical mangrove ecosystems, pp 173–224. American Geophysical Union, Washington, DC

Robertson AI, Daniel PA (1989a) The influence of crabs on litter processing in high intertidal mangrove forests in tropical Australia. Oecologia 78:191–198

Robertson AI, Daniel PA (1989b) Decomposition and the annual flux of detritus from fallen timber in tropical mangrove forests. Limnol Oceanogr 34:640–646

Robertson AI, Dixon P (1993) Separating live and dead roots using colloidal silica: An example from mangrove forests. Plant Soil 157:151–154

Robertson AI, Duke NC (1987) Insect herbivory on mangrove leaves in North Queensland. Aust J Ecol 12:1–7

Robertson AI, Phillips MJ (1995) Mangroves as filters of shrimp pond effluent: predictions and biogeochemical research needs. Hydrobiologia 295:311–321

Robertson AI, Daniel PA, Dixon P (1991) Mangrove forest structure and productivity in the Fly River estuary, Papua New Guinea. Mar Biol 111:147–155

Robertson AI, Alongi DM, Boto KG (1992a) Food chains and carbon fluxes. In: Robertson AI, Alongi DM (eds) Tropical mangrove ecosystems, pp 293–326. American Geophysical Union, Washington, DC

Robertson AI, Daniel PA, Dixon P, Alongi DM (1992b) Pelagic biological processes along a salinity gradient in the Fly delta and adjacent river plume (Papua New Guinea). Cont Shelf Res 13:205–224

Rogerson A, Gwaltney G (2000) High numbers of naked amoebae in the planktonic waters of a mangrove stand in southern Florida, USA J Euk Microbiol 47:235–241

Rogerson A, Anderson OR, Vogel C (2003) Are planktonic naked amoebae predominantly floc associated or free in the water column? J Plankton Res 25:1359–1365

Rojas A, Holguin G, Glick BR, Bashan Y (2001) Synergism between *Phyllobacterium* sp. (N_2-fixer) and *Bacillus licheniformis* (P-solubilizer), both from a semi-arid mangrove rhizosphere. FEMS Microbiol Ecol 35:181–191

Romero LM, Smith III TJ, Fourqurean JW (2005) Changes in mass and nutrient content of wood during decomposition in a south Florida mangrove forest. J Ecol 93:618–631

Rönnbäck P (1999) The ecological basis for economic value of seafood production supported by mangrove ecosystems. Ecol Econ 29:235–252

Rönnbäck P, Macia, A, Almqvist, Schultz, L, Troell M (2002) Do penaeid shrimps have a preference for mangrove habitats? Distribution pattern analysis on Inhaca Island, Mozambique. Est Coast Shelf Sci 55:427–436

Ross MS, Ruiz PL, Telesnicki GJ, Meeder JF (2001) Estimating above-ground biomass and production in mangrove communities of Biscayne National Park, Florida (U.S.A.). Wetlands Ecol Manag 9:27–37

Roth I (1992) Leaf structure: coastal vegetation and mangroves of Venezuela: encyclopedia of plant anatomy, vol 14, Part 2. Gebruder Borntraeger, Berlin

Rowe JS (1994) A new paradigm for forestry. For Chronicle 70:565–568

Roy J, Salager J (1992) Midday depression of net CO_2 exchange of leaves of an emergent tree in French Guiana. J Trop Ecol 8:499–504

Ruitenbeek HJ (1994) Modelling economy-ecology linkages in mangroves: Economic evidence for promoting conservation in Bintuni Bay, Indonesia. Ecol Econ 10:223–247

Saad S, Husain ML, Yaacob R, Asano T (1999) Sediment accretion and variability of sedimentological characteristics of a tropical estuarine mangrove: Kemaman, Terengganu, Malaysia. Mangr Salt Marsh 3:51–58

Saberi BO (1992) The growth and biomass production of *Rhizophora mucronata* in response to the varying concentrations of nitrogen and phosphorus in water culture. Trop Ecol 33:164–171

Saenger P (1982) Morphological, anatomical and reproductive adaptations of Australian mangroves. In: Clough BF (ed) Mangrove ecosystems in Australia, pp 153–191. Australian National University Press, Canberra

Saenger P (2002) Mangrove ecology, siliviculture and conservation. Kluwer, Dordrecht

Saenger P, Snedaker SC (1993) Pantropical trends in mangrove above-ground biomass and annual litterfall. Oecologica 96:293–299

Saintilan N (2004) Relationships between estuarine geomorphology, wetland extent and fish landings in New South Wales estuaries. Est Coast Shel Sci 61:591–601

Saintilan N, Griffiths K, Jaafar W, Tibbey M (2000) A possible experimental artifact associated with leaf-tethering in crab herbivory experiments. Wetlands Aust 18:55–59

Sam R, Ridd PV (1998) Spatial variations of groundwater salinity in a mangrove-salt flat system, Cocoa Creek, Australia. Mangr Salt Marsh 2:121–132

Sánchez BG (2005) Belowground productivity of mangrove forests in southwest Florida. PhD dissertation, Louisiana State University, Baton Rouge, LA

Sasekumar A, Chong VC (1987) Mangroves and prawns: Further perspectives. In: Sasekumar A, Phang SM, Chong VC (eds), Proceedings of the 10th Annual Seminar Malaysian Society of Marine Science, pp 10–22. University of Malaya, Kuala Lumpur

Sasekumar A, Chong VC (eds) (2005) Ecology of Klang Strait. University of Malaya Press, Kuala Lumpur

Saur E, Imbert D, Etienne J, Mian D (1999) Insect herbivory on mangrove leaves in Guadeloupe: Effects on biomass and mineral content. Hydrobiologia 413:89–93

Scholander PF, Van Dam L, Scholander SI (1955) Gas exchange in the roots of mangroves. Am J Bot 42:92–98

Schories D, Barletta-Bergan A, Barletta M, Krumme U, Mehlig U, Rademaker V (2003) The keystone role of leaf-removing crabs in mangrove forests of North Brazil. Wetlands Ecol Manage 11:243–255

Schrijvers J, Vincx M (1997) Cage experiments in an East African mangrove forest: A synthesis. J Sea Res 38:123–133

Schultze E-D (1994) Flux control in biological systems: from enzymes to populations and ecosystems. Academic, San Diego, CA

Scully NM, Maie N, Dailey SK, Boyer JN, Jones RD, Jaffé R (2004) Early diagenesis of plant-derived dissolved organic matter along a wetland, mangrove, estuary ecotone. Limnol Oceanogr 49:1667–1978

Scurlock JMO, Olson RJ (2002) Terrestrial net primary productivity—a brief history and a new worldwide database. Environ Rev 10:91–109

Seitzinger SP (1988) Denitrification in freshwater and coastal marine ecosystems: Ecological and geochemical significance. Limnol Oceanogr 33:702–724

Sengupta A, Chaudhuri S (1991) Ecology of heterotrophic dinitrogen fixation in the rhizosphere of mangrove plant community at the Ganges river estuary in India. Oecologia 87:560–564

Sevrin-Reyssac J (1980) Chlorophyll *a* et production primaire dans les eaux de le baie du Levrier et le Parc National du Banc d'Arguin (Sept–Nov 1974). Bull Cent Nat Rech Oceanogr Pech Nouadhibon (R.I. Mauritainie) 9:56–65

Seymour JR, Seuront L, Mitchell JG (2007) Microscale gradients of planktonic microbial communities above the sediment surface in a mangrove estuary. Est Coast Shelf Sci 73:651–666

Sheaves M (2005) Nature and consequences of biological connectivity in mangrove systems. Mar Ecol Prog Ser 302:293–305

Sheaves M, Molony B (2000) Short-circuit in the mangrove food chain. Mar Ecol Prog Ser 199:97–109

Sherman RE, Fahey TJ, Howarth RW (1998) Soil-plant interactions in a neotropical mangrove forest: Iron, phosphorus and sulfur dynamics. Oecologia 115:553–563

Sherman RE, Fahey TJ, Martinez P (2003) Spatial patterns of biomass and aboveground net primary productivity in a mangrove ecosystem in the Dominican Republic. Ecosystems 6:384–398

Shaiful AAA (1987) Nitrate reduction in mangrove swamps. Malaysian Appl Biol 16:361–367

Shaiful AAA, Abdul Manan DM, Ramli MR, Veerasamy R (1986) Ammonification and nitrification in wet mangrove soils. Malaysian J Sci 8:47–56

Silva CAR, Lacerda LD, Rezende CE (1990) Metals reservoir in a red mangrove forest. Biotropica 22:339–345

Silva CAR, Oliveira SR, Rêgo RDP, Mozeto AA (2007) Dynamics of phosphorus and nitrogen through litter fall and decomposition in a tropical mangrove forest. Mar Environ Res 64: 524–534

Singh HR, Chong VC, Sasekumar A, Lim KH (1994) Value of mangroves as nursery and feeding grounds. In: Wilkinson CR, Surapol S, Chou LM (eds) Proceedings of the third ASEAN-Australia symposium on living marine resources, vol 1, pp 105–122. Chulalongkorn University, Bangkok, Thailand

Skov MW, Hartnoll RG (2002) Paradoxical selective feeding on a low-nutrient diet: Why do mangrove crabs eat leaves? Oecologia 131:1–7

Slim FJ, Hemminga MA, Ochieng C, Jannink NT, Cocheret de la Morinière, van der Velde G (1997) Leaf litter removal by the snail *Terebralia palustris* (Linnaeus) and sesarmid crabs in an East African forest (Gazi Bay, Kenya). J Exp Mar Biol Ecol 215:35–48

Smil V (2008) Energy in nature and society. MIT Press, Cambridge, MA

Smith JAC, Popp M, Lüttge U, Cram WJ, Diaz M, Griffiths H, Lee HSL, Medina E, Schäfer C, Stimmel K-H, Thonke B (1989) Ecophysiology of xerophytic and halophytic vegetation of a coastal alluvial plain in northern Venezuela. VI. Water relations and gas exchange of mangroves. New Phytol 111:293–307

Smith III TJ (1987a) Seed predation in relation to tree dominance and distribution in mangrove forests. Ecology 68:266–273

Smith III TJ (1987b) Effects of light and intertidal position on seedling survival and growth in tropical, tidal forests. J Exp Mar Biol Ecol 110:133–146

Smith III TJ (1992) Forest structure. In: Robertson AI, Alongi DM (eds) Tropical mangrove ecosystems, pp 101–136. American Geophysical Union, Washington, DC

Smith III TJ, Chan H-T, McIvor CC, Robblee MB (1989) Comparisons of seed predation in tropical, tidal forests on three continents. Ecology 70:146–151

Smith III TJ, Boto KG, Frusher SD, Giddens RL (1991) Keystone species and mangrove forest dynamics: The influence of burrowing by crabs on soil nutrient status and forest productivity. Est Coast Shelf Sci 33:419–432

Smoak JM, Patchineelam SR (1999) Sediment mixing and accumulation in a mangrove ecosystem: Evidence from ^{210}Pb, ^{234}Th and ^{7}Be. Mangr Salt Marsh 3:17–27

Snedaker SC (1995) Mangroves and climate change in the Florida and Caribbean region: Scenarios and hypotheses. Hydrobiologia 295:43–49

Snedaker SC, Araújo RJ (1998) Stomatal conductance and gas exchange in four species of Caribbean mangroves exposed to ambient and increased CO_2. Mar Freshw Res 49:325–327

Sobrado MA (2000) Relation of water transport to leaf gas exchange properties in three mangrove species. Trees 14:258–262

Sotomayor D, Corredor JE, Morell JM (1994) Methane flux from mangrove sediments along the southwestern coast of Puerto Rico. Estuaries 17:140–147

Sousa WP, Mitchell BJ (1999) The effect of seed predators on plant distributions: Is there a general pattern in mangroves? Oikos 86:55–66

Spackman W, Scholl DW, Taft WH (1964) Environments of coal formation in Southern Florida. Geological Society of America, Washington, DC

Spackman W, Dolsen CP, Riegel W (1966) Phytogenic organic sediments and sedimentary environments in the Everglades-Mangrove (part I). Paleogeogr Bd 117:135–152

Spain AV, Holt JA (1980) The elemental status of the foliage and branchwood of seven mangrove species from northern Queensland. Division of Soils, Divisional Report No. 49. CSIRO, Canberra

Spalding M, Blasco F, Field C (eds) (1997) World mangrove atlas. Int Soc Mangrove Ecosystems, Okinawa

Spratt Jr HG, Hodson RE (1994) The effect of changing water chemistry on rates of manganese oxidation in surface sediments of a temperate salmarsh and a tropical mangrove estuary. Est Coast Shelf Sci 38:119–135

Stanley SO, Boto KG, Alongi DM, Gillan FT (1987) Composition and bacterial utilization of free amino acids in tropical mangrove sediments. Mar Chem 22:13–30

Staples DJ, Vance DJ, Heales DS (1985) Habitat requirements of juvenile penaeid prawns and their relationship to offshore fisheries. In: Second Aust Natl Prawn Seminar, pp 47–54. CSIRO, Kooralbyn, Queensland

Steinke TD, Ward CJ, Rajh A (1995) Forest structure and biomass of mangroves in the Mgeni estuary, South Africa. Hydrobiologia 295:159–166

Stieglitz T, Ridd PV (2001) Trapping of mangrove propagules due to density-driven secondary circulation in the Normanby River estuary, NE Australia. Mar Ecol Progr Ser 211: 131–142

Stieglitz T, Ridd PV, Hollins S (2000a) A small sensor for detecting animal burrows and monitoring burrow water conductivity. Wetlands Ecol Manag 8:1–7

Stieglitz T, Ridd PV, Muller P (2000b) Passive irrigation and functional morphology of crustacean burrows in a tropical mangrove swamp. Hydrobiologia 421:69–76

Strangmann A, Bashan Y, Giani L (2008) Methane in pristine and impaired mangrove soils and its possible effect on establishment of mangrove seedlings. Biol Fertil Soils 44:511–519

Strom SL (2008) Microbial ecology of ocean biogeochemistry: A community perspective. Science 320:1043–1045

Struve J, Falconer RA, Wu Y (2003) Influence of model mangrove trees on the hydrodynamics in a flume. Est Coast Shelf Sci 58:163–171

Sukardjo S (1995) Structure, litterfall and net primary production in the mangrove forests in East Kalimantan. In: Box EO, Fujiwara T (eds) Vegetation science in forestry, pp 585–611. Kluwer, Dordrecht

Sukardjo S, Yamada I (1992) Biomass and productivity of a *Rhizophora mucronata* Lamarck plantation in Tritih, Central Java, Indonesia. For Ecol Manag 49:195–209

Susilo A, Ridd PV (2005) The bulk hydraulic conductivity of mangrove soil perforated with animal burrows. Wetlands Ecol Manag 13:123–133

Susilo A, Ridd PV, Thomas S (2005) Comparison between tidally driven groundwater flow and flushing of animal burrows in tropical mangrove swamps. Wetlands Ecol Manag 13: 377–388

Suwa R, Khan MNI, Hagihara A (2006) Canopy photosynthesis, canopy respiration and surplus production in a subtropical mangrove *Kandelia candel* forest, Okinawa Island, Japan. Mar Ecol Prog Ser 320:131–139

Swanborough PW, Doley D, Keenan RJ, Yates DJ (1998) Photosynthetic characteristics of *Flindersia brayeyana* and *Castanospermum australe* from tropical lowland and upland sites. Tree Physiol 18:341–347

Tam NFY, Wong YS (1996) Retention of wastewater-borne nitrogen and phosphorus in mangrove soils. Environ Technol 17:851–859

Tanaka K, Choo P-S (2000) Influence of nutrient outwelling from the mangrove swamp on the distribution of phytoplankton in the Matang mangrove estuary, Malaysia. J Oceanogr 56: 69–78

Tanaka N, Sasaki Y, Mowjood MIM, Jinadasa KBSN, Homchuen S (2007) Coastal vegetation structures and their function in tsunami protection: Experience of the recent Indian Ocean tsunami. Landscape Ecol Eng 3:33–45

Thelaus J, Haecky P, Forsman M, Andersson A (2008) Predation pressure on bacteria increases along aquatic productivity gradients. Aquat Micro Ecol 52:45–55

Theuri MM, Kinyamario JI, Speybroeck DV (1999) Photosynthesis and related physiological processes in two mangrove species, *Rhizophora mucronata* and *Ceriops tagal* at Gazi Bay, Kenya. Afr J Ecol 37:180–193

Thibodeau FR, Nickerson NH (1986) Differential oxidation of mangrove substrate by *Avicennia germinans* and *Rhizophora mangle*. Am J Bot 73:513–516

Thomas CR, Christian RR (2001) Comparison of nitrogen cycling in salt marsh zones related to sea-level rise. Mar Ecol Prog Ser 221:1–16

Thomas G, Fernandez TV (1997) Incidence of heavy metals in the mangrove flora and sediments in Kerala, India. Hydrobiologia 352:77–87

Thongtham N, Kristensen E (2003) Physical and chemical characteristics of mangrove crab (*Neoepisesarma versicolor*) burrows in the Bangrong mangrove forest, Phuket, Thailand; with emphasis on behavioral response to changing environmental conditions. Vie Milieu 53: 141–151

Thongtham N, Kristensen E (2005) Carbon and nitrogen balance of leaf-eating sesarmid crabs (*Neoepisesarma versicolor*) offered different food sources. Est Coast Shelf Sci 65:213–222

Tomlinson PB (1986) The botany of mangroves. Cambridge University Press, Cambridge

Torréton J-P, Guiral D, Arfi R (1989) Bacterioplankton biomass and production during destratification in a monomictic eutrophic bay of a tropical lagoon. Mar Ecol Prog Ser 57:53–67

Tremblay LB, Benner R (2006) Microbial contributions to N-immobilization and organic matter preservation in decaying plant detritus. Geochim Cosmochim Acta 70:133–146

Tremblay LB, Dittmar T, Marshall AG, Cooper WJ, Cooper WT (2007) Molecular characterization of dissolved organic matter in a North Brazilian mangrove porewater and mangrove-fringed estuaries by ultrahigh resolution Fourier Transform-Ion Cyclotron Resonance mass spectrometry and excitation/emission spectroscopy. Mar Chem 105:15–29

Tuffers AV, Naidoo G, von Will Cambridge DJ (1999) The contribution of leaf angle to photoprotection in the mangroves *Avicennia marina* (Forssk) Vierh and *Bruguiera gymnorrhiza* (L) Lam under field conditions in South Africa. Flora 194:267–275

Turner IM (2001) The ecology of trees in the tropical rain forest. Cambridge University Press, Cambridge

Turner JT (2004) The importance of small planktonic copepods and their roles in pelagic food webs. Zool Stud 432:255–266

Turner RE (1977) Intertidal vegetation and commercial yields of penaeid shrimp. Trans Am Fish Soc 106:441–456

Twilley RR (1985a) An analysis of mangrove forests along the Gambia River Estuary: Implications for the management of estuarine resources. Great Lakes Mar Wat Cent Int Prog Report No 6, pp 1–75. University of Michigan, Michigan

Twilley RR (1985b) The exchange of organic carbon in basin mangrove forests in a southwest Florida estuary. Est Coast Shelf Sci 20:543–557

Twilley RR (1988) Coupling of mangroves to the productivity of estuarine and coastal waters. In: Jansson B-O (ed) Coastal-offshore ecosystem interactions, pp 155–180. Springer, Berlin

Twilley RR, Chen RH, Hargis T (1992) Carbon sinks in mangroves and their implications to carbon budget of tropical coastal ecosystems. Water Air Soil Pollut 64:265–288

Twilley RR, Gottfried RR, Rivera-Monroy VH, Zhang W, Armijos MM, Bodero A (1998) An approach and preliminary model of integrating ecological and economic constraints of environmental quality in the Guayas River estuary, Ecuador. Environ Sci Policy 1:271–288

Ulanowicz RE, Kay KK (1991) A computer package for the analysis of ecosystem flow networks. Environ Software 6:131–142

Upstill-Goddard RC, Barnes J, Ramesh R (2007) Are mangroves a source or sink for greenhouse gases? In: Tateda Y (ed) Greenhouse gas and carbon balances in mangrove coastal ecosystems, pp 127–138. Gendai Tosho, Kanagawa, Japan

Valentine-Rose L, Layman CA, Arrington DA, Rypel AL (2007) Habitat fragmentation decreases fish secondary production in Bahamian tidal creeks. Bull Mar Sci 80:863–877

Valiela I, Teal JM (1979) The nitrogen budget of a salt marsh ecosystem. Nature 280:652–656

Van der Valk AG, Attiwill PM (1984) Decomposition of leaf and root litter of *Avicennia marina* at Westernport Bay, Victoria, Australia. Aquat Bot 18:205–221

Vazquez P, Holguin G, Puente ME, Lopez-Cortes A, Basham Y (2000) Phosphate-solubilizing microorganisms associated with the rhizosphere of mangroves in a semiarid coastal lagoon. Biol Fertil Soils 30:460–468

Vega-Cendejas ME, Arreguin-Sánchez F (2001) Energy fluxes in a mangrove ecosystem from a coastal lagoon in Yucatan Peninsula, Mexico. Ecol Model 137:119–133

Verma A, Subramanian V, Ramesh R (1999) Day-time variation in methane emission from two tropical urban wetlands in Chennai, Tamil Nadu, India. Curr Sci 76:1020–1022

Verheyden A, Kairo JG, Beeckman H, Koedam N (2004) Growth rings, growth ring formation and age determination in the mangrove *Rhizophora mucronata*. Ann Bot 94:59–66

Verheyden A, Ridder FD, Schmitz N, Beeckman H, Koedam N (2005) High-resolution time series of vessel density in Kenyan mangrove trees reveal link with climate. New Phytol 167: 425–435

Vitousek PM, Sanford Jr RL (1986) Nutrient cycling in moist tropical forests. Annu Rev Ecol Syst 17:137–167

Vo-Luong P, Massel S (2008) Energy dissipation in non-uniform mangrove forests of arbitrary depth. J Mar Syst 74:603–622

Wafar S, Untawale AG, Wafar M (1997) Litter fall and energy flux in a mangrove ecosystem. Est Coast Shelf Sci 44:111–124

Walsh JP, Nittrouer CA (2004) Mangrove-bank sedimentation in a mesotidal environment with large sediment supply, Gulf of Papua. Mar Geol 208:225–248

Wallace AR (1878) Tropical nature and other essays. Macmillan, London

Warburton K (1979) Growth and production of some important species of fish in a Mexican coastal lagoon system. J Fish Biol 14:449–464

Wardle DA, Walker LR, Bardgett RD (2004) Ecosystem properties and forest decline in contrasting long-term chronosequences. Science 305:509–513

Watson JG (1928) Mangrove forests of the Malay Peninsula. Malayan Forest Records No. 6. Federated Malay States, Kuala Lumpur

Wattayakorn G, Wolanski E, Kjerfve B (1990) Mixing, trapping and outwelling in the Klong Ngao mangrove swamp, Thailand. Est Coast Shelf Sci 31:667–688

Wattayakorn G, Prapong P, Noichareon D (2001) Biogeochemical budgets and processes in Bandon Bay, Suratthami, Thailand. J Sea Res 46:133–142

Werry J, Lee SY (2005) Grapsid crabs mediate link between mangrove litter production and estuarine planktonic food chains. Mar Ecol Prog Ser 293:165–176

Whitney DM, Chalmers AG, Haines EB, Hanson RB, Pomeroy LR, Sherr B (1981) The cycles of nitrogen and phosphorus. In: Pomeroy LR, Wiegert RG (eds) The ecology of a salt marsh, pp 163–181. Springer, New York

Wolanski E (1986) An evaporation-driven salinity maximum zone in Australian tropical estuaries. Est Coast Shelf Sci 22:415–424

Wolanski, E (1989) Measurements and modelling of the water circulation in mangrove swamps. UNESCO/COMARAF regional project for research and training on coastal marine systems in Africa-RAF/87/038. Serie Documentarie No. 3, 1–43

Wolanski E (1992) Mangrove hydrodynamics. In: Robertson AI, Alongi DM (eds) Tropical mangrove ecosystems, pp 43–62. American Geophysical Union, Washington, DC

Wolanski, E (1995) Transport of sediment in mangrove swamps. Hydrobiologia 295:31–42

Wolanski E (2007) Estuarine ecohydrology. Elsevier, Amsterdam

Wolanski E, Cassagne B (2000) Salinity intrusion and rice farming in the mangrove-fringed Konkoure River delta, Guinea. Wetlands Ecol Manag 8:29–36

Wolanski E, Gibbs RJ (1995) Flocculation of suspended sediment in the Fly River estuary, Papua New Guinea. J Coast Res 11:754–762

Wolanski E, Ridd P (1986) Tidal mixing and trapping in mangrove swamps. Est Coast Shelf Sci 23:759–771

Wolanski E, Jones M, Bunt JS (1980) Hydrodynamics of a tidal creek-mangrove swamp system. Aust J Mar Freshw Res 31:431–450

Wolanski E, Chappell J, Ridd PV, Vertessy R (1988) Fluidisation of mud in estuaries. J Geophys Res 93:2351–2361

Wolanksi E, Mazda Y, King B, Gay S (1990) Dynamics, flushing and trapping in Hinchinbrook Channel, a giant mangrove swamp, Australia. Est Coast Shelf Sci 31:555–579

Wolanski E, Gibbs RJ, Spagnol S, King B, Brunskill GJ (1998) Inorganic sediment budget in the mangrove-fringed Fly River Delta, Papua New Guinea. Mangr Salt Marsh 2:85–98

Wolanski E, Spagnol S, Thomas S, Moore K, Alongi DM, Trott LA, Davidson A (2000) Modeling and visualizing the fate of shrimp pond effluent in a mangrove-fringed tidal creek. Est Coast Shelf Sci 50:85–97

Wolanski E, McKinnon AD, Williams D, Alongi DM (2006) An estuarine ecohydrology model of Darwin Harbour, Australia. In: Wolanski E (ed) The environment of Asia Pacific harbours, pp 477–488. Springer, Dordrecht

Wolff M, Koch V, Isaac V (2000) A trophic flow model of the Caeté mangrove estuary (North Brazil) with considerations for the sustainable use of its resources. Est Coast Shelf Sci 50: 789–803

Woodroffe CD (1981) Mangrove swamp stratigraphy and Holocene transgression, Grand Cayman Island, West Indies. Mar Geol 41:271–294

Woodroffe CD (1985a) Studies of mangrove basin, Tuff Crater, New Zealand. I. Mangrove biomass and production of detritus. Est Coast Shelf Sci 20: 265–280

Woodroffe CD (1985b) Studies of mangrove basin, Tuff Crater, New Zealand. II. Comparison of volumetric and velocity-area methods of estimating tidal flux. Est Coast Shelf Sci 20: 431–445

Woodroffe CD (1992) Mangrove sediments and geomorphology. In: Robertson AI, Alongi DM (eds) Tropical mangrove ecosystems, pp 7–41. American Geophysical Union, Washington, DC

Woodroffe CD (2003) Coasts: form, process and evolution. Cambridge University Press, Cambridge

Woodroffe CD, Bardsley KN, Ward PJ, Hanley JR (1988) Production of mangrove litter in a macrotidal embayment, Darwin Harbour, N.T., Australia. Est Coast Shelf Sci 26: 581–598

Wösten JHM, deWilligen P, Tri NH, Lien TV, Smith SV (2003) Nutrient dynamics in mangrove areas of the Red River Estuary in Vietnam. Est Coast Shelf Sci 57:65–72

Yabuki K, Kitaya Y, Sugi J (1985) Studies on the function of the mangrove pneumatophore. In: Sugi J (ed) Studies on the mangrove ecosystem, pp 76–79. Tokyo Institute of Agriculture, Toyko

Yamada I (1997) Tropical rain forests of Southeast Asia. University of Hawaii Press, Honolulu, HI

Yánez-Arancibia A, Chavez GS, Sanchez-Gil P (1985) Ecology of control mechanisms of natural fish production in the coastal zone. In: Yánez-Arancibia A (ed) Fish community ecology in estuaries and coastal lagoons: towards an ecosystem integration, pp 571–594. UNAM Press, Mexico

Yates EJ, Ashwath N, Midmore DJ (2002) Responses to nitrogen, phosphorus, potassium and sodium chloride by three mangrove species in pot culture. Trees 16:120–125

Ye Y, Lu C, Lin P (1999) Seasonal and spatial changes in CH_4 emissions from mangrove wetlands of the Hainan Island and Xiamen. Chin J Atmos Sci 23:303–310

Youssef T (1995) Ecophysiology of waterlogging in mangroves. Ph.D. dissertation. Southern Cross University, Lismore, NSW

Youssef T, Saenger P (1996) Anatomical adaptive strategies and rhizosphere oxidation in mangrove seedlings. Aust J Bot 44:297–313

Youssef T, Saenger P (1998) Photosynthetic gas exchange and accumulation of phytotoxins in mangrove seedlings in response to soil physico-chemical characteristics associated with waterlogging. Tree Physiol 18:317–324

Yu K-F, Zhao J-X, Liu T-S, Wang P-X, Qian J-L (2004) Alpha-cellulose $\delta^{13}C$ variation in mangrove tree rings correlates well with annual sea level trend between 1982 and 1999. Geophys Res Lett 31:L11203, doi: 10.1029/200GL019450

Zalewski M, Janauer GA, Jolankaj G (1997) Ecohydrology: A new paradigm for the sustainable use of aquatic resources. In: Conceptual background, working hypothesis, rationale and scientific guidelines for the implementation of the IHP-V projects 2.3/2.4, pp 1–56. Technical documents in Hydrology No. 7.UNESCO, Paris

Zotz G, Harris G, Königer M, Winter K (1995) High rates of photosynthesis in the tropical pioneer tree, *Ficus insipida* Willd. Flora 190:265–272

Index

Lightning Source UK Ltd.
Milton Keynes UK
16 September 2009

143760UK00009B/73/P

9 781402 042706